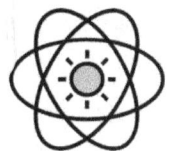

SAT

Physics Subject Test

2017 edition

(with online help)

In Remembrance of

King Bhumibol Adulyadej

About Author:

Ankur Sharma is an experienced professor who had been teaching math and physics for more than fifteen years at AIMS located in Thailand. He earned a B.Engineering a M.SC in TechnologyManagement from Assumption College with full scholarship. Most of the students who studied SAT Physics and Math had pass the exam easily with a score of 650 above

His goal for writing this book was to help students and examinee to use the time efficiently thru the right material with a proper method.

©Copyright 2016 by iGen

All rights reserved

No part of this may be reproduced or distributed without the written permission of the copy

right owner.

For more inquiries please email: ker999@gmail.com

ISBN-13: 978-1540621092

ISBN-10: 154062109X

BISAC: Education / Adult & Continuing Education

CONTENTS

Introduction 8
    Topics on exam 8
    How to use this book 10
    Tips and tricks 11

Chapter 1 - Understanding physics 12
    Fundamental vs Derived units 12
    Units conversion 13
    Graphical analysis 15
    Practice 1 18
    Answer key 1 20

Chapter 2 - Vectors 21
    Vector vs Scalar 21
    Vector operations 22
    Vector components 25
    Practice2 27
    Answer key 2 30

Chapter 3 - Kinematic (Motion) 32
    One dimensional movement 32
    Graphical analysis of motion 34
    Equations of motion 39
    Practice 3 40
    Answer key 3 43

Chapter 4 - Kinematic (Projectiles) 45

- Projectiles 45
- Linking graphs 48
- Circular motion 49
- Practice 4 51
- Answer key 4 55

Chapter 5 - Forces 57

- What is a force 57
- Type of force 58
- Newton's law of motion 60
- Hooke's law 63
- Terminal velocity 65
- Practice 5 66
- Answer key 5 70

Chapter 6 - Work & Energy 72

- Work 72
- Type of energy 74
- Power 78
- Practice 6 79
- Answer key 6 82

Chapter 7 - Momentum & Torque 84

- Momentum 84
- Torque 86
- Angular momentum 89

Practice 7 91

Answer key 7 94

Chapter 8 - Gas & Pressure 96

Pressure 96

Kinetic theory 99

Gas Law 102

Practice 8 104

Answer key 8 107

Chapter 9 - Heat 109

Heat capacity 109

3 ways heat travel 112

Law of thermodynamics 113

Practice 9 116

Answer key 9 119

Chapter 10 - Wave I 121

Type of waves 121

Wave properties 122

Electromagnetic waves 124

Mirrors and Lens 128

Practice 10 132

Answer key 10 137

Chapter 11 - Wave II 139

Simple harmonic motion 139

Standing waves 141

Illumination & polarization 142

Doppler effect 143

Interference of waves & double slit 145

Practice 11 147

Answer key 11 151

Chapter 12 - Gravitational & Electric field 153

Kepler's law 153

Newton's law of gravitational 154

Gravitational potential energy 155

Electric field and potential 158

Practice 12 160

Answer key 12 163

Chapter 13 - Electricity 165

Ohm's law 165

Circuit symbol 167

AC and DC 169

Series and Parallel circuit 170

Other components 172

Practice 13 176

Answer key 13 180

Chapter 14 - Electromagnetic Induction ... 182

    Magnetic field ... 182

    Magnetic force and flux ... 186

    Faraday's law ... 187

    Lenz's law ... 187

    Fleming's rule ... 188

    Generator and transformer ... 189

    Practice 14 ... 192

    Answer key 14 ... 196

Chapter 15 - Quantum physics ... 198

    Inside atom ... 198

    Rutherford model of atom ... 200

    Photoelectric effect & Bohr's model ... 200

    Half-life and Radioactivity ... 205

    Practice 15 ... 210

    Answer key 15 ... 213

Chapter 16 - Special Relativity ... 215

    Law of special relativity ... 215

    Relativity of time, length and mass ... 216

    Relativistic energy ... 218

    Practice 16 ... 219

    Answer key 16 ... 220

Practice Test 223

Answer key 237

Keywords 244

Appendix I 250

Appendix II 251

Appendix III 256

Topics on exam

Mechanics (about 36% to 42%)

- **Kinematics**, such as velocity, acceleration, motion in one dimension, and motion of projectiles
- **Dynamics**, such as force, Newton's laws, statics, and friction
- **Energy and momentum**, such as potential and kinetic energy, work, power, impulse, and conservation laws
- **Circular motion**, such as uniform circular motion and centripetal force
- **Simple harmonic motion**, such as mass on a spring and the pendulum
- **Gravity**, such as the law of gravitation, orbits, and Kepler's laws

Electricity and magnetism (about 18% to 24%)

- **Electric fields, forces, and potentials**, such as Coulomb's law, induced charge, field and potential of groups of point charges, and charged particles in electric fields
- **Capacitance**, such as parallel-plate capacitors and time-varying behavior in charging/discharging
- **Circuit elements and DC circuits**, such as resistors, light bulbs, series and parallel networks, Ohm's law, and Joule's law
- **Magnetism**, such as permanent magnets, fields caused by currents, particles in magnetic fields, Faraday's law, and Lenz's law

Waves and optics (about 15% to 19%)

- **General wave properties**, such as wave speed, frequency, wavelength, superposition, standing wave diffraction, and Doppler effect
- **Reflection and refraction**, such as Snell's law and changes in wavelength and speed
- **Ray optics**, such as image formation using pinholes, mirrors, and lenses
- **Physical optics**, such as single-slit diffraction, double-slit interference, polarization, and color

Heat and thermodynamics (about 6% to 11%)

- **Thermal properties**, such as temperature, heat transfer, specific and latent heats, and thermal expansions
- **Laws of thermodynamics**, such as first and second laws, internal energy, entropy, and heat engine efficiency

Modern physics (about 6% to 11%)

- **Quantum phenomena**, such as photons and photoelectric effect
- **Atomic**, such as the Rutherford and Bohr models, atomic energy levels, and atomic spectra
- **Nuclear and particle physics**, such as radioactivity, nuclear reactions, and fundamental particles
- **Relativity**, such as time dilation, length contraction, and mass-energy equivalence

Miscellaneous (about 4% to 9%)

- **General**, such as history of physics and general questions that overlap several major topics
- **Analytical skills**, such as graphical analysis, measurement, and math skills
- **Contemporary physics**, such as astrophysics, superconductivity, and chaos theory

credit:

https://collegereadiness.collegeboard.org/sat-subject-tests/subjects/science/physics

How to use this book ?

- This book is divided into 17 chapters of full revision

- The formulas and theory are given along with the examples

- Read through the book once before doing the end of chapter questions

- Do the end of chapter questions and check the answer

- This book contain online help with a QR code on a selected question on each chapter

- The QR code will link you to the online resources that will make it easy to do the question

SAT Physics subject test requires you to have at least one year of introduction to pre-college physics and basic knowledge of algebra and trigonometry. It is also advisable that you have some experience in the laboratory experiment.

You will be tested upon fundamental concepts and knowledge, single-concept problem, and mixed concept problem.

Remember that all the formulas require you to convert everything to SI-base units before using them to obtain a correct result.

Tips and Tricks

About the test:

i. No calculator is allowed

ii. There are 75 questions multiple choice

iii. 60 minutes is the time limit

iv. 1 correct gives 1 point

v. 1 wrong gets minus 1/4 point

vi. A blank gives 0 point

vii. Total full score is 800

Guide-line in taking the test:

i. Time yourself

ii. Do easy question first

iii. Each question is awarded same mark

iv. Pace yourself after 5-6 questions

v. Round the numbers for easy calculation

vi. Double check your work for units

vii. Write the formula out

viii. Draw out the diagram for visualization

ix. Eliminate choices

x. Be confident in yourself

How to calculate the score ?

Raw score = number of correct answers - (0.25 x number of wrong answers)

The raw score is then converted using a curved calculation for overall score into range of 200 to 800.

Understanding Physics

1

Check-list:

- ✓ Fundament Units VS Derived Units
- ✓ Units Conversion
- ✓ Graphical Analysis

What is Physics ?

Physics is a science of understanding and applying concept of energy efficiently and effectively. Three main elements in physics are:

1. Theory → Understanding the concept and the formula

2. Calculation → Knowing which formula to use in supporting the theory

3. Experiment → Verifying the theory and calculation graphically or verbally

Fundament Units VS Derived Units

SI-base units are ,also known as International System of Units, used widely by scientist to support the measurement taken during the experiments. It is easy to remember that **fundamental unit** as a measured unit during experiment, while **derived unit** as a mixture of units during calculation.

Table 1.1 Fundamental Units

Quantity	Unit	Symbol Used
Length	meter	m
Mass	kilogram	kg
Time	second	s
Temperature	kelvin	k
Electric Current	ampere	A
Amount of substance	mole	mol

SAT Physics (Physics made simple)

Derived Units Example

$$\text{Speed (m/s)} = \frac{\text{Distance (m)}}{\text{Time (s)}} \leftarrow \text{FUNDAMENTAL UNITS}$$

↑ DERIVED UNITS

Table 1.2 Derived Units

Quantity	Unit	Symbol Used	Fundamental Unit
Speed	Meters per second	m/s	m/s
Area	Meters square	m^2	m^2
Volume	Meters cube	m^3	m^3
Force	Newton	N	kg·m/s^2
Pressure	Pascal	Pa	kg/m·s^2
Electric Charge	Coulomb	C	A·s

Units Conversion

Table 1.3 Units Conversion

Prefixes	Value	Standard form	Symbol
Tera	1,000,000,000,000	10^{12}	T
Giga	1,000,000,000	10^9	G
Mega	1,000,000	10^6	M
Kilo	1,000	10^3	k
deci	0.1	10^{-1}	d
centi	0.01	10^{-2}	c
milli	0.001	10^{-3}	m
micro	0.000001	10^{-6}	μ
nano	0.000000001	10^{-9}	n
pico	0.000000000001	10^{-12}	p

Example 1.1

Convert 5234 ns to seconds

Step1: look for nano standard form → 10^{-9}

Step2: rewrite 5234 ns → 5234 x 10^{-9}

= 0.000005234 *seconds*

Example 1.2

Convert 6.9392 Mm to meter

Step1: look for Mega standard form → 10^6

Step2: rewrite 6.9392 Mm → 6.9392 x 10^6

= 6,939,200 *meters*

Example 1.3

Convert 24 cm^2 to m^2

Step1: look for centi standard form → 10^{-2}

Step2: rewrite 24 cm^2 → 24 x (10^{-2} x 10^{-2})

= 0.0024 m^2

We have to multiply twice by 10^{-2} because unit of area was squared

Graphical Analysis

When an experiment is performed the result are normally shown by plotting graph, to demonstrate the relationship between the variables. We can categorize the relationships as followed:

1. **Direct Relationship** (or directly proportional)

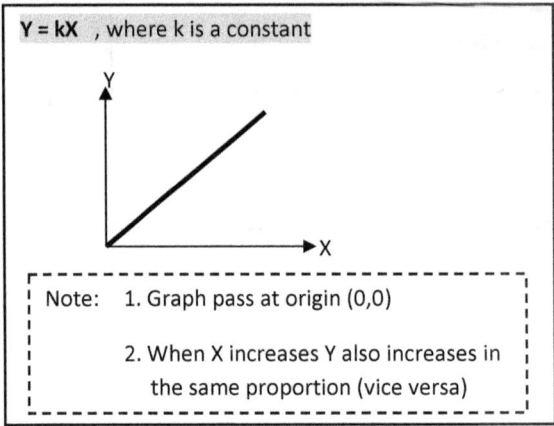

Y = kX , where k is a constant

Note: 1. Graph pass at origin (0,0)

2. When X increases Y also increases in the same proportion (vice versa)

2. **Indirect Relationship** (or inversely proportional)

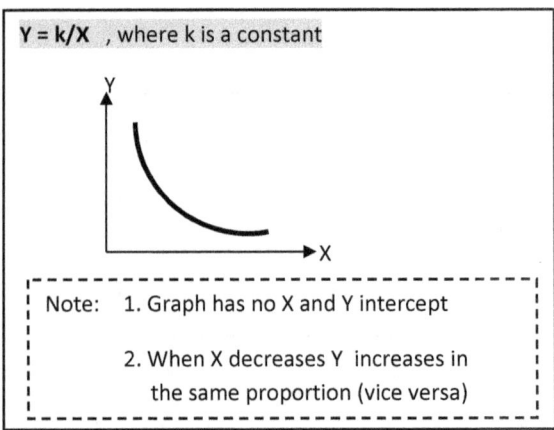

Y = k/X , where k is a constant

Note: 1. Graph has no X and Y intercept

2. When X decreases Y increases in the same proportion (vice versa)

3. **Constant Relationship**

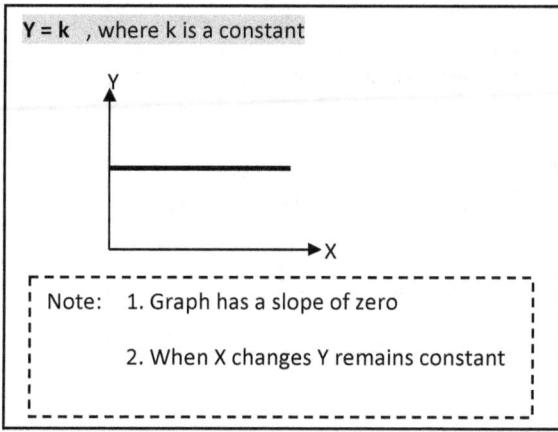

4. **Square Relationship**

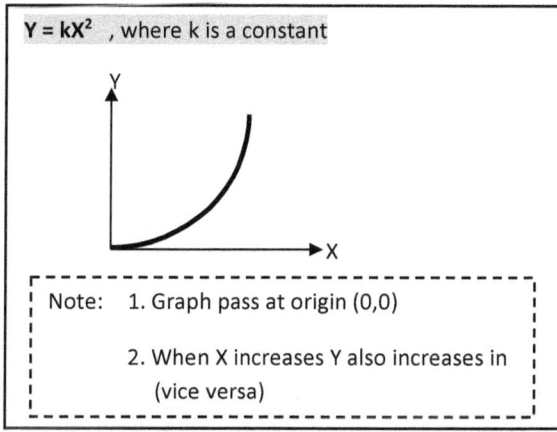

5. **Square-root Relationship**

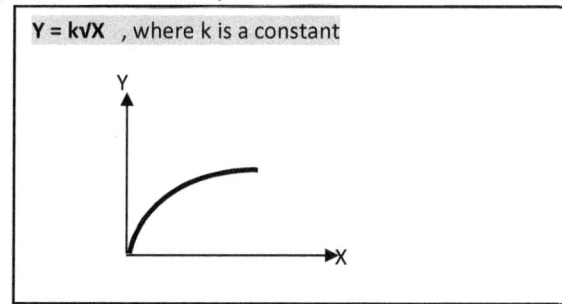

Example 1.4

What is the relationship between kinetic energy of an object and its velocity?

Solution: We first need to know the formula to link 'kinetic energy' and 'velocity' of an object → $KE = 1/2\ mv^2$ (from here we know that velocity is being squared)

Therefore, the relationship is a square relationship or $KE = kv^2$

Practice Chapter 1

1. Which of the following is a derived units ?

    a) Kelvin

    b) Ampere

    c) Kilogram

    d) Candela

    e) Newton

2. A spiral bacteria has a length of 2.647 μm, this is equivalent to

    a) 0.00002647 m

    b) 0.000002647 mm

    c) 0.0002647 cm

    d) 0.02647 mm

    e) 0.0000002647 m

3. Distance from the Sun to the Earth is measured in AU(Astronomical Units). 1 AU is equal to 150 Million Kilometer, how would this be written in Standard form ?

    a) 150×10^6 Km

    b) 1.5×10^8 Km

    c) 1.5×10^{11} Km

    d) 15×10^{11} Mm

    e) 1.5×10^9 m

4. Convert 230 mm² to m²

   a) 0.00023 m²

   b) 0.023 m²

   c) 0.23 m²

   d) 2.3 m²

   e) 230000 m²

5. Which of the graph shows that 'Y' is inversely proportional to the square root of 'X' ?

   a)

   b)

   c)

   d)

   e)

SAT Physics (Physics made simple)

Answer: Chapter 1

Question	Answer	Explanation
1.	E	Kelvin, Ampere, Kilogram and Candela are all base units The only derived unit here is Newton
2.	C	2.647 µm is mathematically equivalent to 0.0002647 cm
3.	B	150 Million Kilometer can be written as $1.50 \times 10^2 \times 10^6$ Km = 1.5×10^8 Km
4.	A	230 mm^2 = 230 $\times (10^{-3})^2$ m^2 = 230 $\times 10^{-6}$ m^2 = 0.000230 m^2
5.	E	'Y' is inversely proportional to the square root of 'X' can be translated as Y = k/√x , whose graph can be plotted similarly to inversely proportional graph with a limitation that X must be greater than zero

SAT Physics (Physics made simple)

Vectors   2

Check-list:

- ✓ Vector VS Scalar
- ✓ Vector Operations
- ✓ Vector components

Vector quantity VS Scalar quantity

In physics we are concern with the sign in front of quantity as well as in math. For instance when we say a car is moving at 50 km/hr, we know that it will cover a distance of 50 km in 1 hour but we don't know which direction is it travelling. So we need the direction to find the proper ending point from the starting point.

Table 2.1 Vector and Scalar

Quantity	Definition	Examples
Scalar	has only magnitude (size)	Distance , Speed , Mass Time, Temperature, Energy
Vector	has magnitude and direction	Displacement, Velocity, Weight Force, Acceleration, Momentum

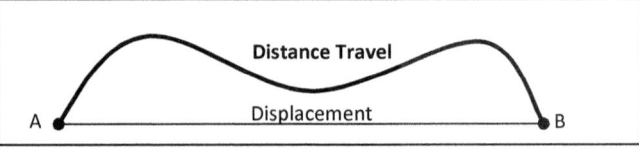

Scalar x Vector	=	Vector
Scalar x Scalar	=	Scalar
Vector x Vector	=	Vector

Example of Vector and Scalar Multiplication

SAT Physics (Physics made simple)

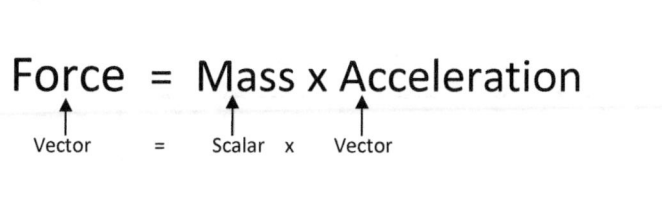

Vector Operation

A vector quantity has a starting point and an ending point with a direction represents by sign or arrow. Here is an example of how a vector(u) can be manipulated:

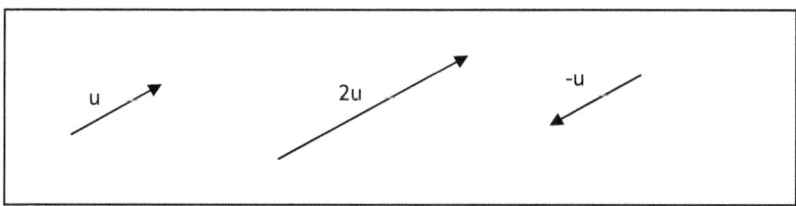

Example 2.1

Given that **c = 2a - b**

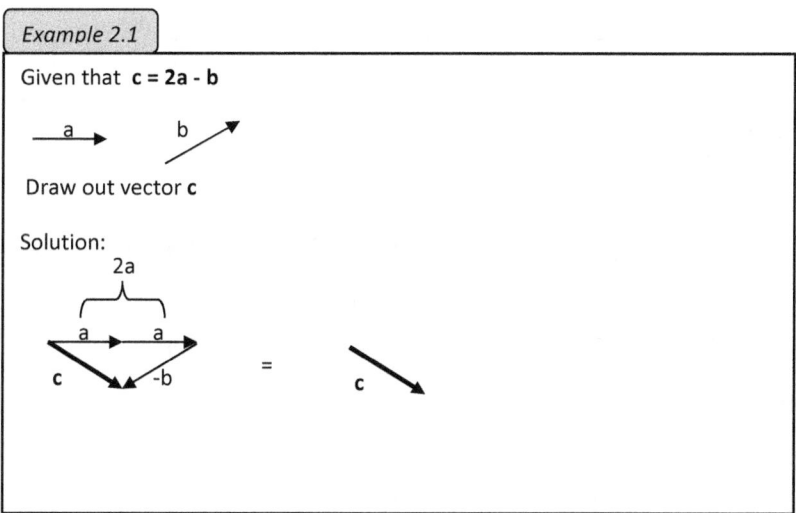

Vector Addition

[5 N →, 8 N →] = [5 + 8 = 13 N Resultant →]

Vector Subtraction

[← 5 N, 8 N →] = [8 - 5 = 3 N Resultant →]

Vector with Pythagoras

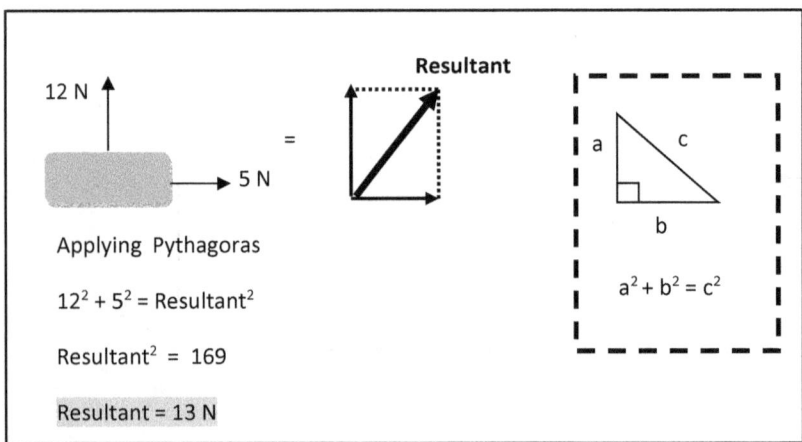

12 N ↑ , 5 N → = Resultant (diagonal)

Applying Pythagoras

$12^2 + 5^2 = Resultant^2$

$Resultant^2 = 169$

Resultant = 13 N

$a^2 + b^2 = c^2$

SAT Physics (Physics made simple)

Vector with parallelogram concept

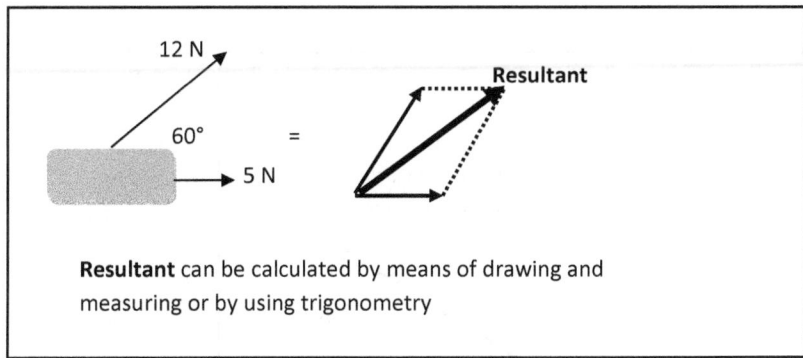

Resultant can be calculated by means of drawing and measuring or by using trigonometry

Rules for trigonometry

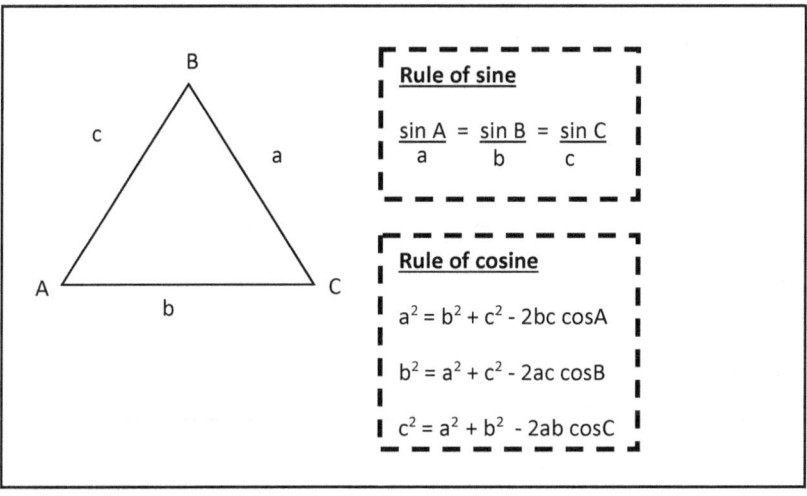

Rule of sine

$$\frac{\sin A}{a} = \frac{\sin B}{b} = \frac{\sin C}{c}$$

Rule of cosine

$a^2 = b^2 + c^2 - 2bc \cos A$

$b^2 = a^2 + c^2 - 2ac \cos B$

$c^2 = a^2 + b^2 - 2ab \cos C$

Example 2.2

Find the resultant of the following diagram

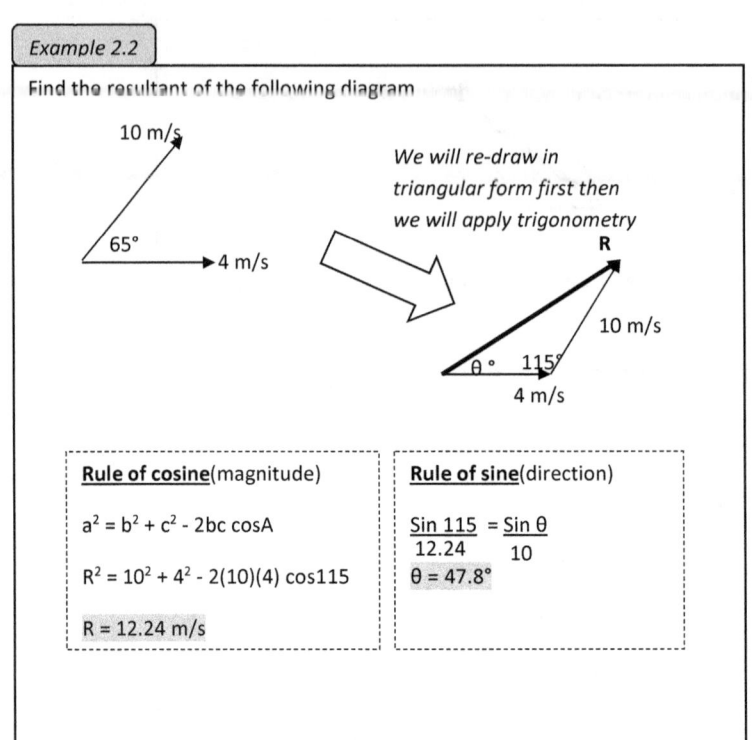

Rule of cosine (magnitude)

$a^2 = b^2 + c^2 - 2bc \cos A$

$R^2 = 10^2 + 4^2 - 2(10)(4) \cos 115$

$R = 12.24 \text{ m/s}$

Rule of sine (direction)

$\dfrac{\sin 115}{12.24} = \dfrac{\sin \theta}{10}$

$\theta = 47.8°$

Vector Component

A vector drawn normally consist of two component (x and y). If we say a vector that is A m and $\theta°$ from horizontal then we are getting this:

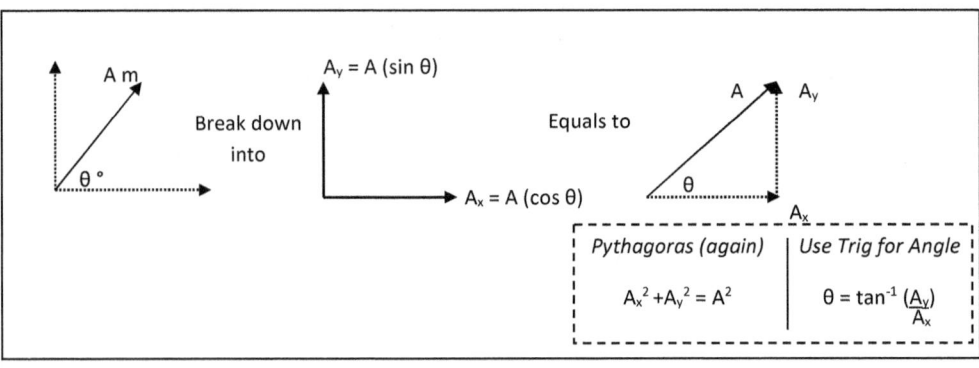

SAT Physics (Physics made simple)

Example 2.3

What are the vertical and horizontal component of a ship moving at 30 m/s with the angle of 30° east of north?

Solution:

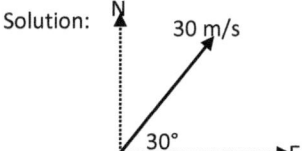

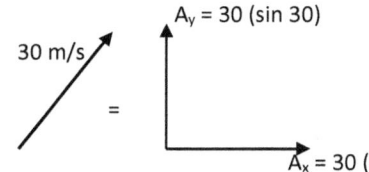

$A_y = 30 (\sin 30)$

$A_x = 30 (\cos 30)$

Vertical component

$A_y = 30 \sin(30) = 15$ m/s

Horizontal component

$A_x = 30 (\cos 30) = 26$ m/s

Practice Chapter 2

1. Which pair contains one scalar and one vector quantity ?

   a) work and force
   b) force and acceleration
   c) mass and speed
   d) distance and time
   e) power and speed

2. Which of the following pair of force gives a resultant of 3 N ?

   a) 2 N and 10 N
   b) 2 N and 13 N
   c) 10 N and 10 N
   d) 10 N and 14 N
   e) 10 N and 23 N

3. What would be the resultant velocity on the boat if it is pulled by a crane 9 m/s due west while the current in the sea is 12m/s due north ?

   a) 15 m/s 36.9 ° North-West
   b) 15 m/s 53.1 ° North-West
   c) 15 m/s 36.9 ° North-East
   d) 21 m/s 36.9 ° North-West
   e) 21 m/s 53.1 ° North-East

4. Which expression gives the vertical component of a 60N force applied to the book at 12° to horizontal ?

   a)   60 cos(12)

   b)   60 sin(12)

   c)   60 tan(12)

   d)   12 cos(60)

   e)   12 sin(60)

5. What is the magnitude of resultant of vector below, given that both have a magnitude of 5 m and angle between them is 120°?

   a)   8.66 m

   b)   7.23 m

   c)   6.81 m

   d)   5.00 m

   e)   4.50 m

6. Which of the following represents the resultant of vector a - 2b?

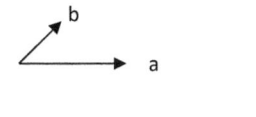

   a)   b)    c)

   d)   e)

7. Given that Vector C has a component of $C_x = 6$ and $C_y = 13$, what angle would C makes with the positive Y-axis

   a) 82.55

   b) 65.22

   c) 44.11

   d) 32.66

   e) 24.77

8. Which of the following represents the resultant of vector operation below ?

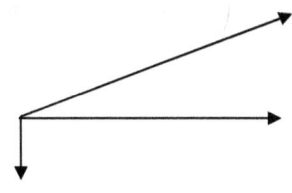

a)

b)

c)

d)

e) ↑

Answer: Chapter 2

Question	Answer	Explanation
1.	A	**Work** has no direction, therefore it is a 'scalar' quantity **Force** has a direction, therefore it is a 'vector' quantity
2.	C	Create a min and max resultant table the resultant should lie between it Minimum obtained by subtracting Maximum obtained by adding 	Pairs	Min	Max
2, 10	8	12
2, 13	11	15
10, 10	0	20	← 3 lies between them | 10, 14	4	24
10, 23	13	33	
3.	a	Find magnitude of resultant (R) by using Pythagoras $R^2 = 9^2 + 12^2$ $\underline{R = 15 \text{ m/s}}$ Find direction using Trig $\theta = \tan^{-1}(9/12) = \underline{36.9° \text{ N of W}}$
4.	b	For vertical component we use $A_y = A \sin\theta$ in this case Vertical component = $\underline{60 \sin(12)}$
5.	D	Since both vector have the same magnitude and the angle between one of the is 60 this makes them become an equilateral triangle. R is therefore equal to 5m

SAT Physics (Physics made simple)

Question	Answer	Explanation
6.	C	vector a - 2b , we draw vector 'a' first then followed by '-2b'
7.	E	The angles that C makes with Y-axis is calculated by using $\tan(\theta) = C_x / C_y$ $\theta = \tan^{-1}(6/13)$ $\underline{\theta = 24.77°}$
8.	A	

Kinematic - Motion   3

Check-list:

- One-dimensional movement
- Graphical analysis of motion
- Equations of motion

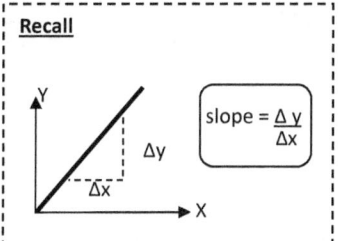

Recall

slope = $\frac{\Delta y}{\Delta x}$

One-dimensional motion

Distance is a measurement of total length travel by an object.

Displacement is a measurement of length, along with direction, from starting point to ending point.

Example 3.1

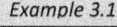

Position	Distance (units)	Displacement (units)
AB	3	- 3
ABC	9	+ 3
ABA	6	0

Speed is a measured of how fast an object moves (rate at which an object travels)

$$\text{Speed (m/s)} = \frac{\text{Change in Distance (m)}}{\text{Change in Time (s)}} = \frac{\Delta \text{Dist}}{\Delta \text{Time}}$$

SAT Physics (Physics made simple)

Velocity is a measured of how fast and which direction an object moves from the beginning to ending (rate at which an object displaces)

$$\text{Velocity (m/s)} = \frac{\text{Change in Displacement (m)}}{\text{Change in Time (s)}} = \frac{\Delta \text{Disp}}{\Delta \text{Time}}$$

Example 3.2

Find the speed and the velocity of the object from A to B below :

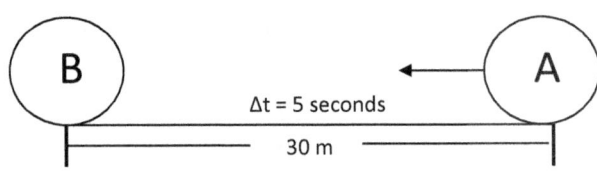

Δt = 5 seconds
30 m

Speed = $\frac{\Delta \text{dist}}{\Delta \text{time}}$ = $\frac{30 \text{ m}}{5 \text{ s}}$

Speed = 6 m/s

Velocity = $\frac{\Delta \text{disp}}{\Delta \text{time}}$ = $\frac{-30 \text{ m}}{5 \text{ s}}$

Velocity = - 6 m/s

Average Speed (m/s) = $\frac{\text{Total Distance (m)}}{\text{Total Time (s)}}$

Average Velocity (m/s) = $\frac{\text{Total Displacement (m)}}{\text{Total Time (s)}}$

SAT Physics (Physics made simple)

Graphical analysis of motion

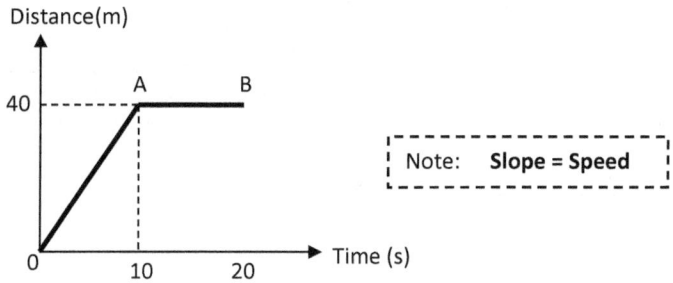

Distance VS Time Graph

Note: **Slope = Speed**

Position	Distance (m)	Speed (m/s)
OA	40	40 ÷ 10 = 4
AB	0	0

Since we know that the object travels a total distance of 40 m and the total time is 20 seconds, we can calculate the 'average speed'

$$\text{Average speed} = \frac{40 \text{ m}}{20 \text{ s}} = 2 \text{ m/s}$$

It's important to understand that the graph of distance vs time does not indicated the direction of motion.

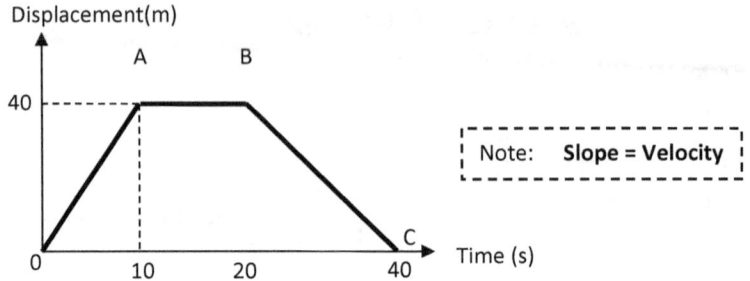

Displacement VS Time Graph

Note: **Slope = Velocity**

Position	Displacement (m)	Velocity (m/s)
OA	40	40 ÷ 10 = + 4 (moving forward)
AB	0	0 (at rest)
BC	- 40	-40 ÷ 20 = - 2 (moving back)

Since we know that the object travels *a total distance of 40 m* and the total time is 40 seconds, we can calculate the 'average speed'

$$\text{Average speed} = \frac{80 \text{ m}}{40 \text{ s}} = 2 \text{ m/s}$$

Now we know that the object travels a *total displacement of 0 m* and the total time is 40 seconds, we can calculate the 'average speed'

$$\text{Average velocity} = \frac{0 \text{ m}}{40 \text{ s}} = 0 \text{ m/s}$$

Example 3.3

The graph below shows variation of a bus moving from one bus stop to another

a) describe the motion on each segment

b) calculate average speed and average velocity

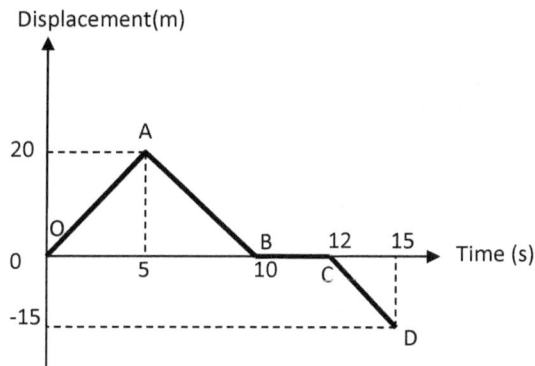

Position	Displacement (m)	Velocity (m/s)
OA	20	$20 \div 5 = +4$ (moving forward)
AB	-20	$-20 \div 5 = -4$ (moving back)
BC	0	0 (at rest)
CD	-15	$-15 \div 3 = -5$ (moving back)

b) Total distance = 20 + 20 + 15 = 55 m, Total time = 15 sec

Average speed = $\frac{55 \text{ m}}{15 \text{ s}}$ = **3.67 m/s**

Total displacement = 20 - 20 - 15 = -15 m, Total time = 15 sec

Average velocity = $\frac{-15 \text{ m}}{15 \text{ s}}$ = **-1 m/s**

Acceleration is a measured of rate at which an object's velocity changes.

$$\text{Acceleration (m/s}^2\text{)} = \frac{\text{Change in Velocity (m/s)}}{\text{Change in Time (s)}} = \frac{\Delta V}{\Delta T} = \frac{V_2 - V_1}{t_2 - t_1}$$

Example 3.4

Find the acceleration of the object from A to B below:

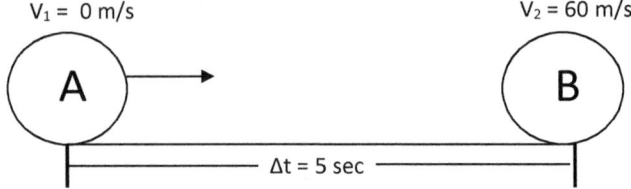

Acceleration $= \frac{\Delta V}{\Delta T} = \frac{60 \text{ m}}{5 \text{ s}} = 12 \text{ m/s}^2$

acceleration = 12 m/s²

This means that every 1 second velocity increases by 12 m/s

Keep in mind that when acceleration is positive the object is speeding up. Vice versa if the acceleration is negative the object is slowing down.

Velocity VS Time Graph

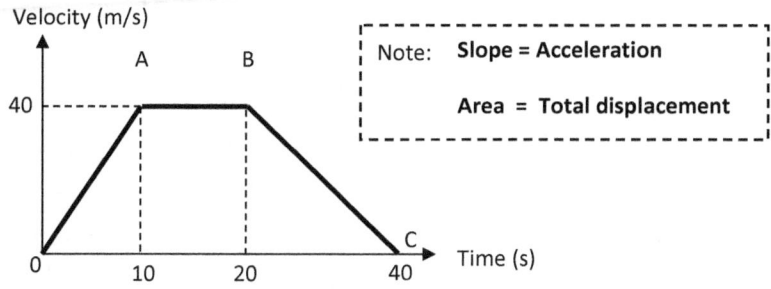

Note: **Slope = Acceleration**

Area = Total displacement

Position	Displacement (m)	Acceleration (m/s²)
OA	1/2 x 10 x 40 = 200	40 ÷ 10 = + 4 (accelerating)
AB	10 x 40 = 400	0 (constant velocity)
BC	1/2 x 20 x 40 = 400	-40 ÷ 20 = - 2 (decelerating)

In this case the total distance is equal to the total displacement, since the graph never come lower than time axis (x-axis) which implies that the object was traveling forward throughout its journey.

Therefore total distance & displacement = 200 + 400 + 400 = 1,000 m

$$\text{Average speed} = \text{Average velocity} = \frac{1000 \text{ m}}{40 \text{ s}} = \underline{25 \text{ m/s}}$$

Equations of motion

There are 4 equations of motion required in calculating the questions regarding movement in 1-D or 2-D, given that there is acceleration in the system.

Below are the equations used:

Equations	Missing term
$v = u + at$	s
$s = ut + 1/2\, at^2$	v
$v^2 = u^2 + 2as$	t
$s = \dfrac{(u+v)\, t}{2}$	a

Variable	Meaning
u	Initial velocity
v	Final velocity
a	Acceleration
t	Time
s	Distance

Example 3.5

A car initially at rest accelerate to velocity of 20 m/s in 4 seconds.

a) what is the total distant traveled ?

solution: List the know variables

$u = 0$ m/s (since the car is initially at rest)
$v = 20$ m/s (final speed of the car)
$t = 4$ s (time taken)
$s = ??$ (missing 'a')

$$s = \frac{(u+v)\,t}{2} = \frac{(0+20) \times 4}{2} = \underline{40\ m}$$

b) what is its acceleration ?

the only missing term would be 's' for this case

so we use: $v = u + at$

$20 = 0 + a(4) \rightarrow a = \underline{5\ m/s^2}$

SAT Physics (Physics made simple)

Practice
Chapter 3

1. Which property of a graph allows acceleration to be determined ?
   a) the area of a velocity-time graph
   b) the area of a displacement-time graph
   c) the slope of a velocity-time graph
   d) the slope of a distance-time graph
   e) the slope of a displacement-time graph

Use below graph to answer question 2 to 4

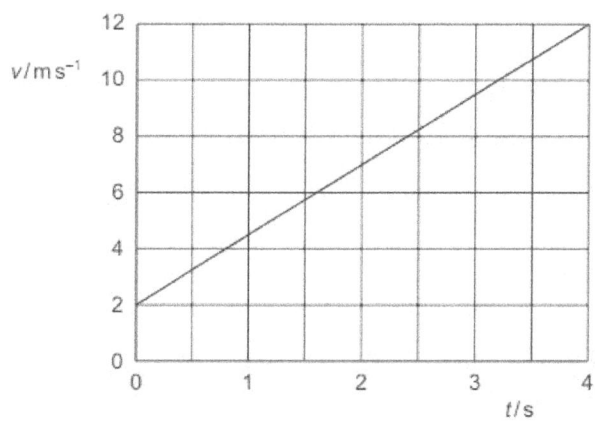

2. What is the distance travel during the four second period?
   a) 8 m
   b) 20 m
   c) 28 m
   d) 32 m
   e) 36 m

3. What is the acceleration of the at 2 second period ?

   a) 2   m/s²

   b) 2.5 m/s²

   c) 3   m/s²

   d) 3.5 m/s²

   e) 4   m/s²

4. What is the average velocity of this object?

   a) 2 m/s

   b) 5 m/s

   c) 6 m/s

   d) 7 m/s

   e) 9 m/s

5. Which of the graph shows an object moving in an object coming to rest ?

   a)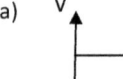

   b)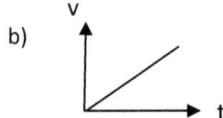

   c)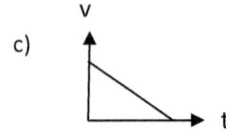

   d) 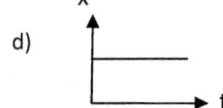   e)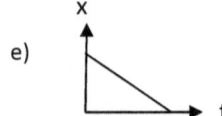

6. New AMG GTS claims that it can do zero to 100 km/hr in 3.5 seconds. What is the approximate acceleration of this sports car ?

   a)   12   m/s²

   b)   10   m/s²

   c)   9.5 m/s²

   d)   9   m/s²

   e)   8   m/s²

7. According to question 6 what would be the distance travel during the acceleration to 100 km/hr of this sports car ?

   a)   48.6 m

   b)   52.3 m

   c)   69.5 m

   d)   89.2 m

   e)   108   m

8. New AMG GTS claims that it's ceramic breaks can stop the car of speed 100 km/hr in about 80 m. What is the approximate time taken to stop ?

   a)   10 s

   b)   9   s

   c)   8   s

   d)   6   s

   e)   5   s

Scan me

9. A ball was thrown up with to the height of 20 m, what is the initial velocity of the throw ?

   a)   30 m/s

   b)   20 m/s

   c)   15 m/s

   d)   10 m/s

   e)   8   m/s

SAT Physics (Physics made simple)

Answer: Chapter 3

Question	Answer	Explanation
1.	C	Slope of velocity-time graph will gives the value of the acceleration Acceleration = $\frac{\Delta velocity}{\Delta time}$
2.	C	The graph is in the shape of trapezium Distance = Area = $\frac{1}{2}(a+b) \times h$ Distance = $\frac{1}{2}(2+12) \times 4 = 28$ m
3.	B	Acceleration = Slope = $\frac{\Delta Velocity}{\Delta time}$ = $\frac{10 \text{ m/s}}{4 \text{ s}}$ = 2.5 m/s²
4.	D	Average speed = $\frac{\text{total distance}}{\text{total time}}$ = $\frac{28 \text{ m}}{4 \text{ s}}$ = 7 m/s
5.	C	An object coming to rest is easily observed from Velocity vs Time graph at first it should have some initial velocity then it should slope down and end at time axis (where v = 0)
6.	E	final velocity = $\frac{100 \text{ km}}{1 \text{ hr}}$ = $\frac{100 \times 1000 \text{ m}}{1 \text{ hr} \times 60\text{min} \times 60 \text{ sec}}$ = 27.78 m/s Acceleration = $\frac{\Delta Velocity}{\Delta time}$ = $\frac{27.78 - 0}{3.5 \text{ s}}$ = 8 m/s²

Question	Answer	Explanation
7.	A	Here we use the equation of motion We know $u = 0$ m/s, $v = 27.78$ m/s, $t = 3.5$ sec, $s = ??$ So we use $s = \dfrac{(u+v) \times t}{2} = \dfrac{(0+27.78) \times 3.5}{2} = 48.6$ m
8.	D	Here we use the equation of motion We know $v = 0$ m/s, $u = 27.78$ m/s, $s = 80$ m, $t = ??$, So we use $s = \dfrac{(u+v) \times t}{2} \rightarrow t = \dfrac{2 \times s}{u} = \dfrac{2 \times 80}{27.7} = 5.76$ sec ≈ 6 sec
9.	B	Here we use the equation of motion We know $v = 0$ m/s, $u = ??$, $s = 20$ m, $a = -10$ m/s^2 So we use $v^2 = u^2 + 2as \rightarrow 0^2 = u^2 + 2(-10)(20) \rightarrow u = 20$ m/s

Kinematic - Projectiles   4

Check-list:

- ✓ Projectiles
- ✓ Linking motion graphs
- ✓ Circular motion

What is a Projectile ?

Projectiles are motion under the influence of x and y component velocity. It's easier to say motion traveling in X-Y coordinate plane parabolic form.

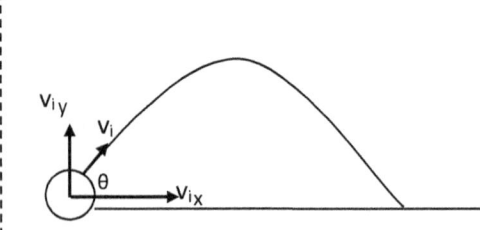

Here we can see the path travel by the object

$V_{ix} = V_i \cos(\theta)$ | $V_i^2 = V^2_{ix} + V^2_{ix}$ | $\tan \theta = \dfrac{V_{iy}}{V_{ix}}$

$V_{iy} = V_i \sin(\theta)$ | $|V_i| = \sqrt{V^2_{ix} + V^2_{ix}}$

Now we can break the motion into X-component and Y-component. In X-component gravitational pull of the Earth has no effect on motion, but in Y-component we have to consider the effect of gravity on acceleration due to gravitational pull which is 9.8 m/s² ≈ 10 m/s⁵

SAT Physics (Physics made simple)

Acceleration due to gravity

Every object on Earth experiences the same gravitational pull of the Earth, $a_g = 9.8$ m/s². Below is an example of how an object drop would have gained the velocity by the pull of the Earth.

A ball drop from rest → $V_{iy} = 0$ m/s   at   time = 0 sec

$a_g = 9.8$ m/s² ≈ +10 m/s²

$V_{iy} = 10$ m/s   at   time = 1 sec

$V_{iy} = 20$ m/s   at   time = 2 sec

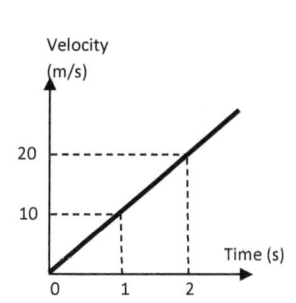

Projectile

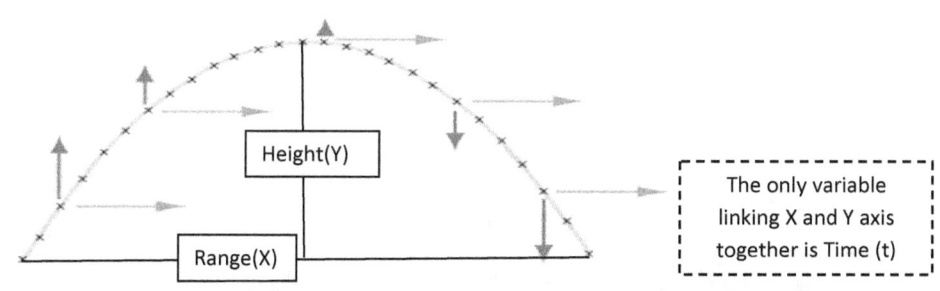

The only variable linking X and Y axis together is Time (t)

X-axis equations	Y-axis equations
Acceleration due to gravity has no effect on the motion in x-axis, therefore a = 0 m/s²	Acceleration due to gravity has an effect on the motion in y-axis, therefore a ≈ 10 m/s²
$V_{ix} = \dfrac{x}{t}$	$Y = V_{iy}t - \dfrac{a_g t^2}{2}$
Range = $V_{ix} \times 2t_{up}$	$t_{up} = \dfrac{V_{iy}}{a_g}$

Example 4.1

A ball is thrown vertically up from the ground with the speed of 30 m/s, how high does it rise and how long does it take to fall back to the ground ? (ignore air resistance)

Solution: In this question there is only Y component velocity

$a = 10$ m/s², $V_{iy} = 30$ m/s, $V_{fy} = 0$ m/s, $t_{up} = $??

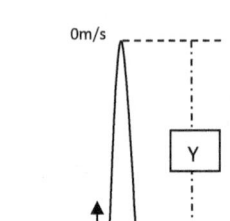

$t_{up} = \dfrac{30}{10} = 3$ sec

Height $Y = $?? , $Y = V_{iy}t - \dfrac{a_g t^2}{2}$

$Y = 30(3) - 0.5 \times 10 \times (3)^2$

$\underline{Y = 45 \text{ m}}$

To find total time, we know it took 3 seconds to go up so it will take 3 seconds to come down, therefore total time will be 3 + 3 = <u>6 seconds</u>

Example 4.2

A ball is dropped vertically down from a cliff 40m , how long does it take to hit ground ? (ignore air resistance)

Solution: In this question there is only Y component velocity

$a = 10$ m/s², $V_{iy} = 0$ m/s, $y = 40$ m, $t = $??

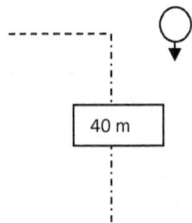

$Y = V_{iy}t - \dfrac{a_g t^2}{2}$

$40 = 0(t) - 0.5 \times 10 \times (t)^2$

$40 = 5 t^2$

<u>t = 2.83 seconds</u>

It will take about 2.83 seconds for the ball to fall

Graphs of projectiles

Let's kick a ball in this curve path

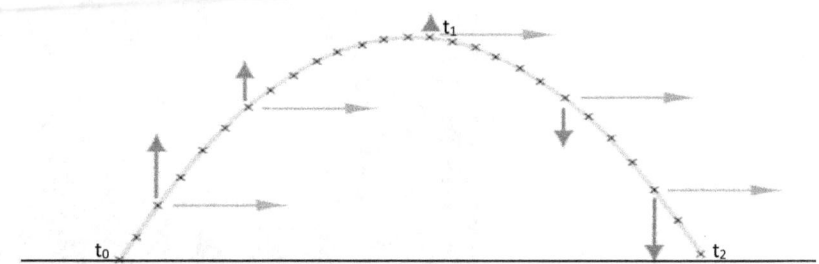

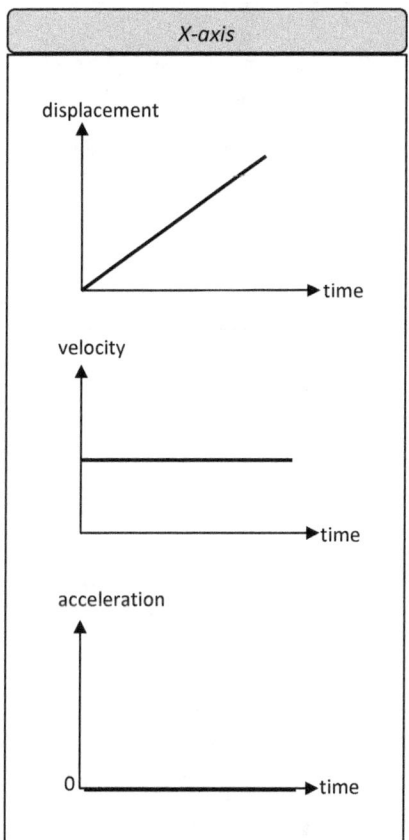

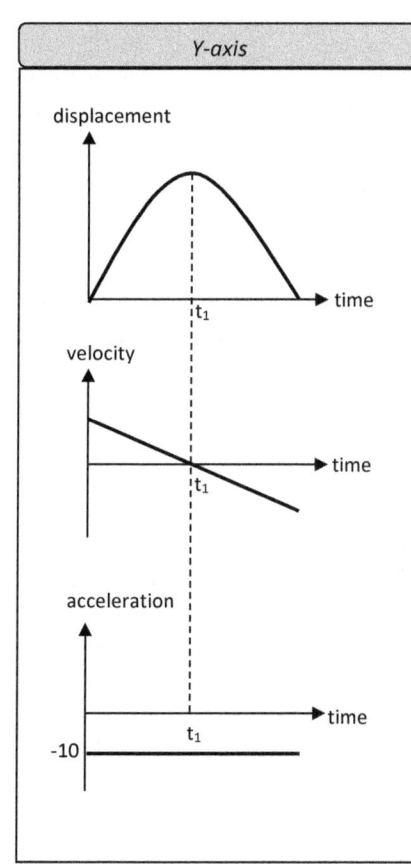

Circular Motion

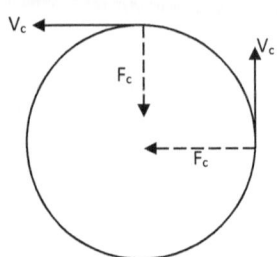
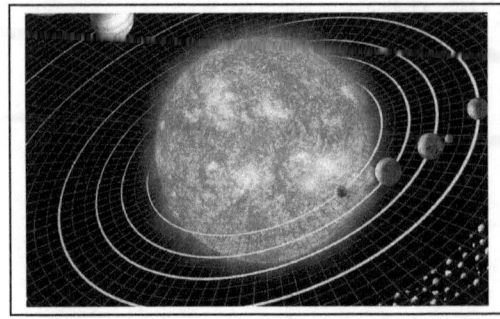

Any object moving in circular path will have a **centripetal force(F_c)**, this force act towards the center of the circle. Once the force is removed the object will move with a **tangential velocity (V_c)**

One complete revolution around a circular path will give a distance equal to circumference of the circle ($2\pi r$). Hence the displacement around a circular path is equal to zero. In circular motion we will take calculation of the angle in terms of radians.

Example 4.3

Convert 60° to radian

solution: $60° \times \dfrac{\pi}{180} = \dfrac{3}{\pi}$ radian or $\dfrac{3}{3.14}$ = 1.047 rad

Convert $\dfrac{\pi}{6}$ to degree

solution: $\dfrac{\pi}{6} \times \dfrac{180°}{\pi} = 30°$

Now all our calculation will be in radians

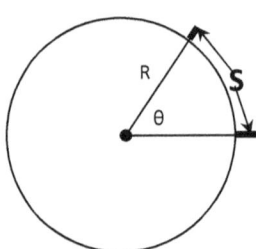

Formula

$S = \theta R$

$V = \dfrac{\Delta S}{\Delta T} = \dfrac{\Delta \theta R}{\Delta T} = \omega R$

$\omega = \dfrac{\Delta \theta}{\Delta T} = 2\pi f$

understanding frequency (f)

$f = \dfrac{1}{period} = \dfrac{1}{T}$ ← 1 Revolution / time taken (Second)

Hertz (Hz)

S - circular distance traveled(m),
V - tangential velocity (m/s),
ω - angular velocity (rad/s)
f - frequency (Hz)

Example 4.4

A disc with a radius of 5 cm spins on a cd player at the rate of 200 rpm.

Calculate a) distance travel in 1 minute b) angular velocity c) tangential velocity

solution: a) distance travel around circle is calculated by $2\pi r$

200 rpm   = 1 minutes → 200 round

Distance travel = $2\pi(5)$ x 200 round = 2000π cm = 2000 x 3.14 cm

= 62.8 meters

b) angular velocity $\omega = 2\pi f$ (we need frequency)

frequency = 200 ÷ 60 (we want frequency Hz or s^{-1})

f = 33.333 Hz

$\omega = 2\pi (33.333) = 209$ rad/s

c) tangential velocity → $V = \omega R = 209 \times 0.05 = 10.5$ m/s

Practice Chapter 4

1. An object thrown vertically up at the velocity of 50 m/s, ignoring air resistance, will take how many seconds to reach maximum height ?

    a)   5

    b)   6

    c)   8

    d)   9.8

    e)   10.5

2. A ball is kicked at the speed of 25m/s with the angle of 60° to the ground, what is its maximum altitude ?

    a)   13.5 m

    b)   23.4 m

    c)   30.5 m

    d)   64.7 m

    e)   71.8 m

3. According to the ball in question2, how long will its flight time be ?

    a)   1.5 s

    b)   1.75 s

    c)   2.2 s

    d)   3.5 s

    e)   4.4 s

4. According to the ball in question2, how far horizontally does the ball travel ?

   a)   23.4 m

   b)   35 m

   c)   55 m

   d)   63 m

   e)   78 m

5. Which of the graph shows of horizontal velocity vs time graph of an object in question 2 ?

   a)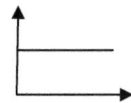

   b)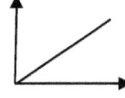

   c)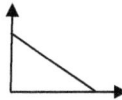

   d)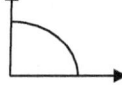

   e)

6. A plane moving at the speed of 350 km/hr dropped a bomb, 4 seconds later it hit the ground. At what altitude was the bomb dropped ?

   a)   40   m

   b)   80   m

   c)   250   m

   d)   400   m

   e)   1400 m

7. Marry-go-round has a radius of 35 m took about 10 second to complete one revolution. What is its angular velocity ?

   a)   0.14 rad/s

   b)   0.24 rad/s

   c)   0.53 rad/s

   d)   0.63 rad/s

   e)   0.77 rad/s

   Scan me

8. According to question7, what is the tangential velocity ?

   a)   3.5 m/s

   b)   9.5 m/s

   c)   18 m/s

   d)   22 m/s

   e)   34 m/s

9. A ball roll off the roof-top horizontally with the speed of 12 m/s, how far from the house will it land if the roof-top and the house is 5 m tall ?

   a)   12 m

   b)   18 m

   c)   24 m

   d)   30 m

   e)   36 m

SAT Physics (Physics made simple)

10. The graph below is the graph of displacement vs time

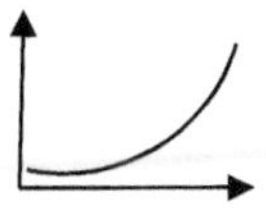

    Which of the following best define the graph characteristic ?

	Velocity	Acceleration
a)	constant	constant
b)	increasing	zero
c)	constant	zero
d)	zero	constant
e)	increasing	constant

11. The graph below is the graph of velocity vs time

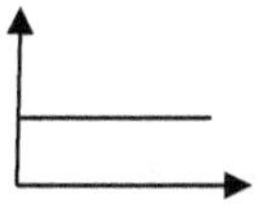

    Which of the following best define the graph characteristic ?

	Displacement	Acceleration
a)	constant	constant
b)	increasing	zero
c)	constant	zero
d)	decreasing	constant
e)	increasing	constant

Answer: Chapter 4

Question	Answer	Explanation
1.	A	$t_{up} = \dfrac{V_{iy}}{g} = \dfrac{50}{10} = 5$ seconds
2.	B	We will first find $V_{iy} = V_i \sin\theta = 25 \sin(60) = 21.6$ m/s Now we use $V_{fy}^2 = V_{iy}^2 - 2a_g S \rightarrow 0 = 21.6^2 - 2(10)S$ $S = 23.4$ m
3.	E	$t_{up} = \dfrac{V_{iy}}{g} = \dfrac{21.6}{10} = 2.2$ seconds if it takes 2.2 second to go up it will take the same time to come down therefore, total flight time = 2.2 + 2.2 = 4.4 seconds
4.	C	Range = $V_{ix} \times 2t_{up}$ = 25 cos(60) × 2(2.2) = 55 m
5.	A	During the entire flight of the projectile the velocity in the x-component remains the same because gravity has no effect on this component $V_{ix} = V_x = V_{fx} = 25 \cos(60)$ Therefore our graph will be a horizontal line above time-axis
6.	B	In this question we only have V_{iy} (V_{ix} = 0) So we use $Y = V_{iy}t - 0.5a_g t^2 \rightarrow Y = 0(4) - 0.5(10)(4)^2 \rightarrow Y = 80$ m
7.	D	$\omega = \dfrac{\Delta\theta}{\Delta t} = \dfrac{2\pi}{10} = 0.63$ rad/s $2\pi \rightarrow 1$ revolution of the wheel
8.	D	$V = \omega R = 0.63 \times 35 = 22$ m/s
9.	A	First find time taken to fall down, we know that V_{iy} is zero Using: $Y = V_{iy}t - 0.5a_g t^2 \rightarrow 5 = 0(t) + 0.5(10)t^2 \rightarrow t = 1$ sec We want distance travel in x-axis now $V_{ix} = \dfrac{x}{t} \rightarrow x = t\, V_{ix} = 1 \times 12 = 12$ m

Question	Answer	Explanation
10.	E	The graph of displacement vs time have an increasing gradient (since it's not a linear graph) so we can say that the velocity will be the gradient of this graph. Therefore velocity is increasing as a linear function. If velocity is linear then acceleration will be constant
11.	B	If the velocity is constant all the time the distance will keep on increasing linearly, without any increase in acceleration

Forces   5

Check-list:

- ✓ What is a force
- ✓ Type of forces
- ✓ Newton Law of motion
- ✓ Hooke's Law

What is a Force ?

Force is a push or a pull, it is a vector quantity. Force can be applied or exerted on to an object.

Force can change object's:

i) Motion and direction
ii) Shape/size
iii) Velocity

An example here shows a man pushing the box with a force F, the box moves along the same direction that the force is applied with a velocity V.

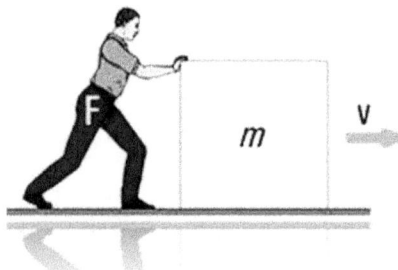

$$\text{Force} = \text{Mass} \times \text{Acceleration}$$
$$(N) \qquad (Kg) \qquad (m/s^2)$$

SAT Physics (Physics made simple)

Type of forces

Weight is a force applied on a body by the pull of the Earth. Gravitational pull on Earth or acceleration due to gravity is 9.81 m/s² or 10 m/s². Remember that mass always remain the same (everywhere) the only change that occur is on weight depending on the value of acceleration due to gravity.

$$\text{Weight} = \text{Mass} \times \text{Acceleration due to gravity}$$
$$\text{(N)} \qquad \text{(Kg)} \qquad (10 \text{ m/s}^2)$$

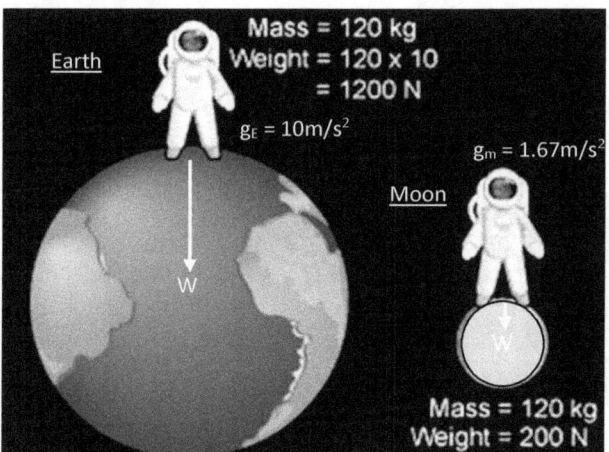

Frictional force is a force that opposes the motion, it always act in the opposite direction of motion. Slowing moving object down or even stopping them.

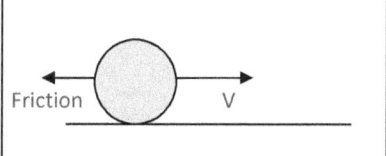

Friction = Coefficient of friction × Normal

$$f = \mu \times N$$
$$\text{(N)} \qquad\qquad \text{(N)}$$

There are two type of friction:

1. Static friction - when object is at rest we need to overcome this force to get it moving

2. Dynamic friction - when the object is moving this type of friction slows it down

Tension is a force created upon by a string or wire attached to a mass or masses.

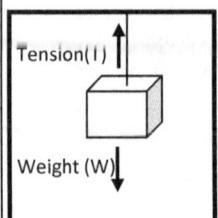

In this case the box is not moving, so we can say that

Tension on the string = Weight of the box

Centripetal force - force that keeps the object moving in circle, this force acts towards the center of the circle. One good example is the force that causes the Earth to orbits around the Sun.

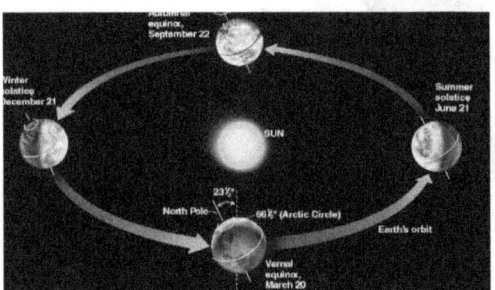
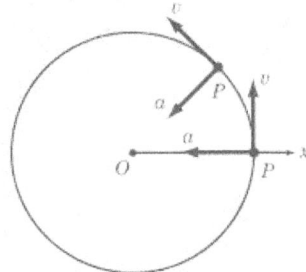

The Sun exerts the centripetal force to keep the Earth and other planets orbiting around it in an elliptical path. The force and acceleration in this case are pointing in the same direction, while the tangential velocity will be perpendicular to the acceleration vector.

Centripetal force = mass x acceleration

$$F_c = m \times a_c$$
$$(N) \quad (kg) \quad (m/s^2)$$

Newton's law of motions

First Law:

Part 1

 →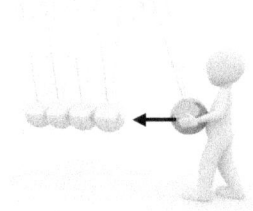

An object at rest remains at rest → *unless acted upon by a net force*

Part 2

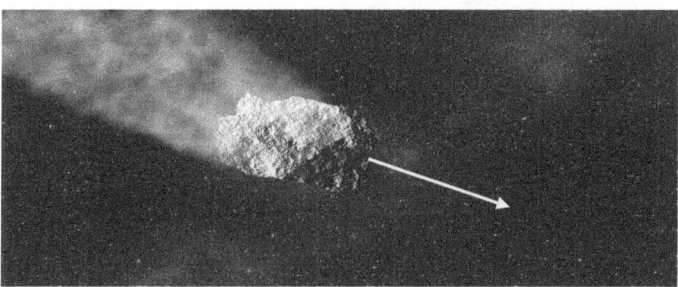

An object in motion remains in motion, unless acted upon by a net force

---

Second Law:

An unbalance force causes an object to accelerate. Force is directly proportional to mass or acceleration.

$$\text{Force} = \text{Mass} \times \text{Acceleration}$$

Third Law:

If an object exerts a force on a second object (action force), the second object in return exerts the same amount of force in the opposite direction (reaction force).

$$\text{Action Force} = \text{Reaction Force}$$

*in the opposite direction

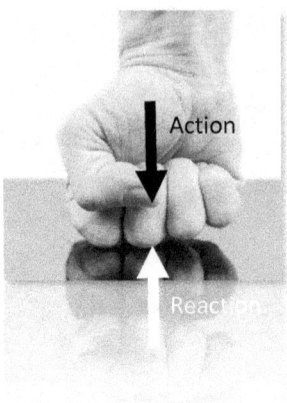

Example 5.1

Draw a free body diagram on a plane moving at a constant velocity

solution:

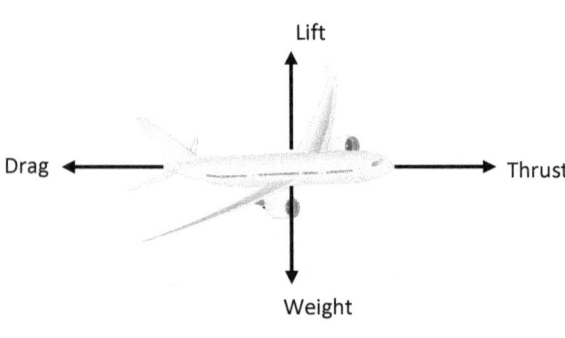

Example 5.2

A ball of mass 0.5 kg is kicked horizontally with a velocity of 15 m/s after 3 seconds the ball comes to rest. Calculate the frictional force experienced by the ball.

solution: First draw out free body diagram

Friction ← [ball] → Kicked Force

Friction = Kicked Force

Friction = m x a = 0.5 x -5 = **-2.5 N**

Frictional force on the ball is 2.5 Newton backward

We will find 'a' by using equation of motion.

u = 15 m/s, v = 0 m/s, t = 3 seconds, a = ??

$$a = \frac{v - u}{t} = \frac{0 - 15}{3} = -5 \text{ m/s}^2$$

Free-body-diagram on incline plane

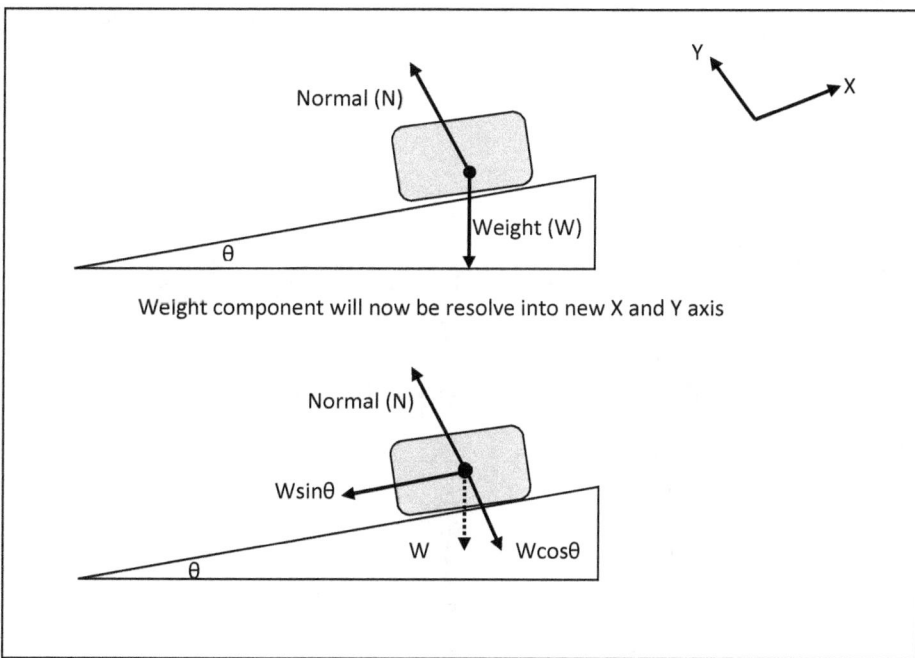

Weight component will now be resolve into new X and Y axis

SAT Physics (Physics made simple)

Hooke's Law

> Extension is proportional to the Load

$$\text{Force} = \text{spring constant} \times \text{extension}$$
$$(N) \qquad\quad (N/m) \qquad\qquad (m)$$

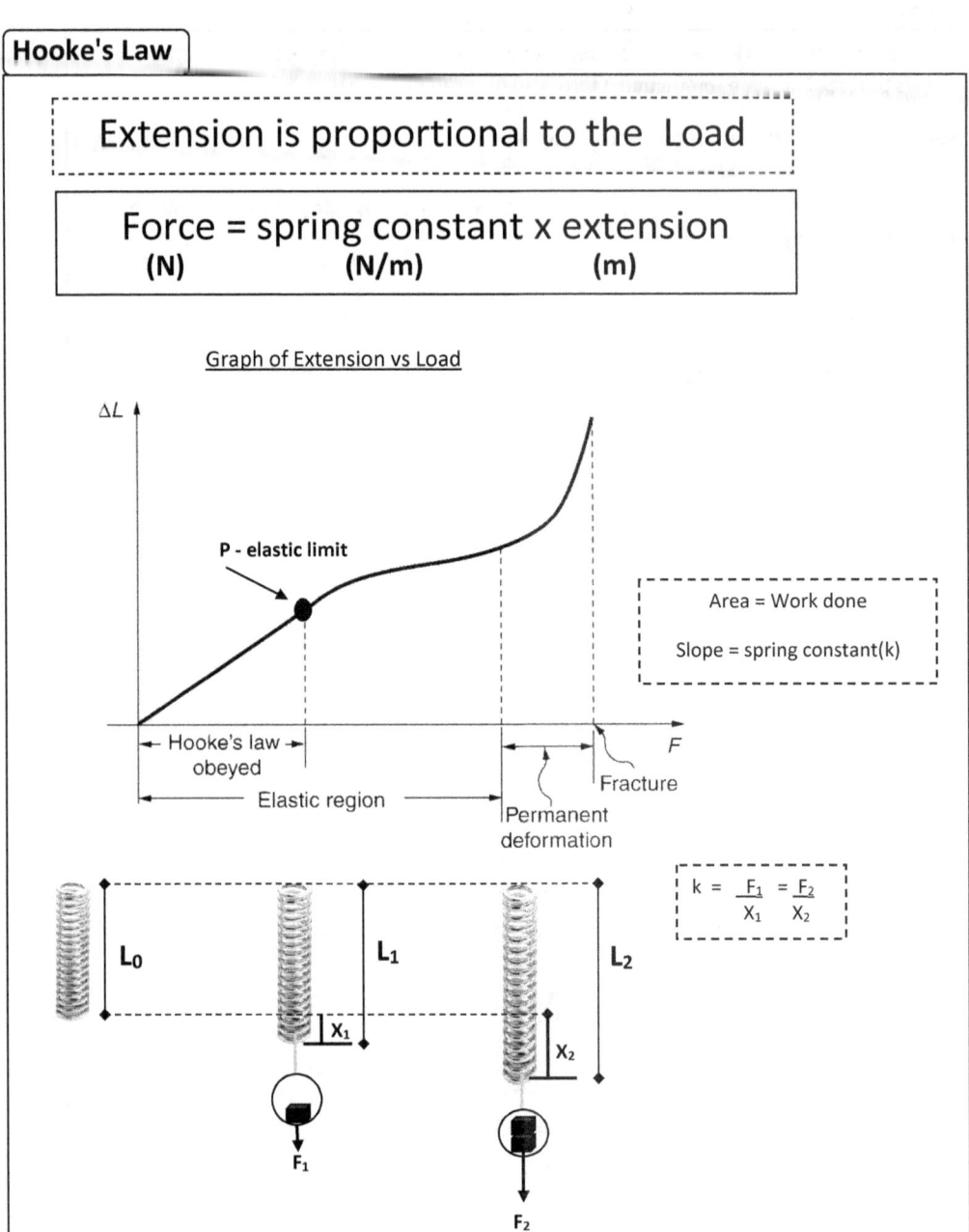

Example 5.3

A 5 cm spring long has a load of 6 N hung on it extend to the length of 6 cm. What is the extension if 18 N load is hang on it ?

Solution: We first find the extension1 → $X_1 = L_1 - L_0 = 6 - 5 = 1$ cm

Now we can take ratio of the load over extension

$$\frac{F_1}{X_1} = \frac{F_2}{X_2} \rightarrow \frac{6 N}{1 cm} = \frac{18 N}{X_2} \rightarrow X_2 = 18/6 = \underline{3\ cm}$$

The extension on the spring will be 3 cm if 18 N is hung on it

Example 5.4

A mass of 1.5 kg is pushed against the spring whose spring constant is 4N/cm. The spring is compressed to the distance of 3cm on a horizontal floor. How much frictional force must the floor exerts to stop the mass ? Find also coefficient of friction

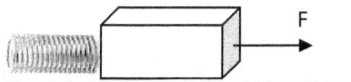

Solution: We know that $F = kx \rightarrow F = 4 \times 3 = 12$ N

We know that Force from spring(F) stop by friction (f) should be equal

So $f = F = \underline{12\ N}$

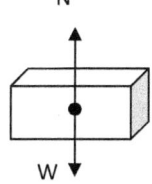

To find μ (coefficient of friction) we need to relate to Free body diagram

$N = W = mg = 1.5 \times 10 = 15$ N → $f = \mu N$ → $12 = \mu \times 15$ → $\mu = \underline{0.8}$

Coefficient of friction in this case is 0.8

Terminal Velocity

Every object on Earth experiences a gravitational pull of the Earth, a_g = 9.8 m/s². At one point the object will start falling at a constant velocity, this occurs when upward force is equal to downward force.

A ball drop from rest

W

V_{iy} = 0 m/s   at time = 0 sec

a_g = 9.8 m/s² ≈ +10 m/s²

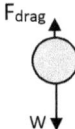
F_{drag}
W

V_{iy} = 10 m/s   at time = 1 sec

W > F_{drag}

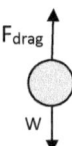
F_{drag}
W

V_{iy} = 20 m/s   at time = 2 sec

W = F_{drag} → F_{net} = 0 N
At this point the acceleration will be zero

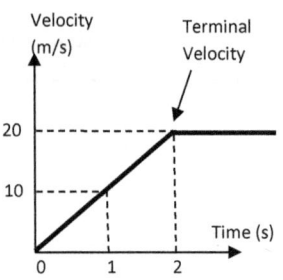

According to the graph object will be moving at a constant velocity after reaching terminal velocity, until it hit the ground. Every object has its own terminal velocity.

SAT Physics (Physics made simple)

Practice
Chapter 5

1. A metal ball with a mass of 0.8kg is pushed horizontally from rest with a speed of 12m/s in 0.2 second by a toy gun. How much force does the gun exerts on the ball ?

   a) 9.6 N

   b) 16.2 N

   c) 28 N

   d) 36 N

   e) 48 N

2. Which planet would the astronaut of mass 60 kg experience a force of 671.4N ?

Planet	Gravitational Strength (m/s^2)
a) Mercury	0.36
b) Venus	0.88
c) Jupiter	26.04
d) Saturn	11.19
e) Uranus	10.49

3. If a plane is gliding at a constant velocity then which of the following must be true ?

   a) Only net force on X-axis is zero

   b) Only net force on Y-axis is zero

   c) Both net force on X and Y axis is zero

   d) Net force on X-axis is increasing

   e) Both net force on X-axis and Y-axis is increasing

4. Find tension on the string 1 ?

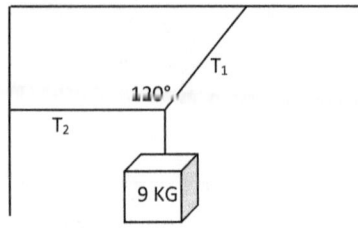

a) 90 N

b) 104 N

c) 155 N

d) 180 N

e) 196 N

5. An object of 5 kg rest on an incline plane that is 4 m away from a 3 m high wall without any movement. What is the frictional force on the object ?

Scan me

a) 30 N

b) 37.5 N

c) 40 N

d) 66 N

e) 83.3 N

6. In question 5, if a man wants the box to accelerate at the rate of 10 m/s^2 then how much force must he exerts on to the box ?

a) 30 N

b) 50 N

c) 60 N

d) 110 N

e) 135 N

7. A 100 g marble lying on the table is pushed against the spring with a spring constant of 50 N/m, if the spring was compressed by 4 cm then how much force is applied on to the marble by the spring ?

   a) 2   N

   b) 3   N

   c) 5   N

   d) 10 N

   e) 13 N

8. According to question 7, what is the acceleration of the marble due to the spring?

   a) 0.5 m/s²

   b) 1.5 m/s²

   c) 14 m/s²

   d) 20 m/s²

   e) 34 m/s²

9. According to question 7, what is the coefficient of friction if the ball comes to rest in 2 second ?

   a) 0.3

   b) 0.48

   c) 0.82

   d) 1.32

   e) 2.00

10. As the angle θ increases what happens to F_x and F_y

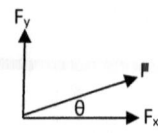

	F_x	F_y
a)	no change	increases
b)	increases	no change
c)	decreasses	increases
d)	decreases	decreases
e)	increases	decreases

Answer: Chapter 5

Question	Answer	Explanation
1.	E	$F = ma = m\dfrac{(v-u)}{t} = 0.8 \times \dfrac{(12-0)}{0.2} = 48$ N
2.	D	We use $W = m \times g \rightarrow 671.4 = 60 \times g \rightarrow g = 11.19$ m/s² Saturn is the only planet that has a gravitational pull of 11.19 m/s²
3.	C	Constant velocity means that both X and Y axis will have a resultant force or net force of zero
4.	B	$T_1\cos 30$ 30° T_1 $W = 90$ N Since the box is at rest we say F_{net} on Y-axis $= 0$ N $F_{up} = F_{down}$ $T_1 \cos 30 = 90$ $T_1 = 104$ N
5.	A	From Pythagoras Normal (N) $W\sin\theta$ f 5m 3 m θ $W = 50$N $W\cos\theta$ 4 m Since the object is not moving we can say that $W\sin\theta = f$ $50(3/5) = f$ We know $\sin\theta = \dfrac{opp}{hyp} = \dfrac{3}{5}$ $f = 30$ N

Question	Answer	Explanation
6.	D	*[Diagram: inclined plane with base 4m, height 3m, hypotenuse 5m, angle θ. Block on incline with W=50N, Wsinθ down the incline, Wcosθ perpendicular, friction f down the incline, applied force F=?? up the incline]* $F_{net} = F - f - W\sinθ$ $ma = F - f - W\sinθ$ $5(10) = F - 30 - 30$ $F = 110$ N
7.	A	*[Diagram: spring attached to wall with ball, force F pulling right]* $F = kx$ $F = 50(0.04)$ $F = 2$ N
8.	D	$F = ma$ $2 = 0.1 \times a$ $a = 20$ m/s²
9.	E	$F = f = 2$ N $2 = μN$ $2 = μ(mg)$ $μ = 2$
10.	C	$F_x = F\cosθ$ $F_y = F\sinθ$ As θ increases Cosθ → decreases Sinθ → increases Since F is constant we can conclude that X-component decreases and Y-component increases

Work & Energy   6

Check-list:

- ✓ Work
- ✓ Type of energy
- ✓ Power

Work

Work done is the amount of energy change take place. If we move an object from point A to point B then a work is done. If no distance is moved then no work is done.

An example here shows a man doing work by pushing the box with a force F.

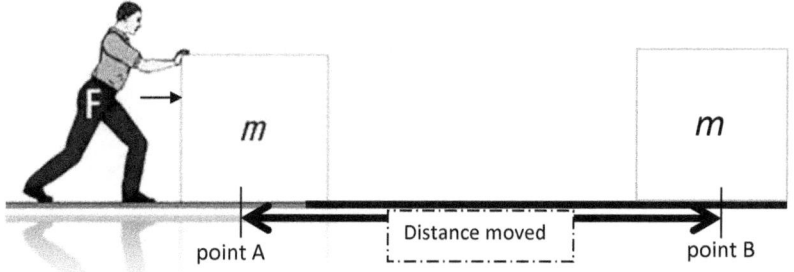

$$\text{Work} = \text{Force} \times \text{Distance}$$
(Joules)   (Newton)   (meter)

$$\text{Work} = \Delta \text{Energy}$$

SAT Physics (Physics made simple)

Example 6.1

A box of mass 3 kg is lifted from the table and put on the shelf 5 meter away. Calculate done by the person lifting the box.

solution:

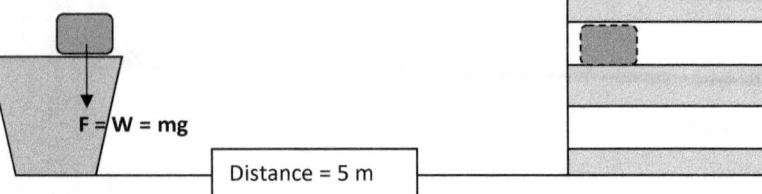

Work = Force x distance

Work = (mg) x d = (3 x 10) x 5 = <u>150 Joules</u>

150 J of energy is transformed during the lifting of the object

Type of energy

Energy is the ability to do work.

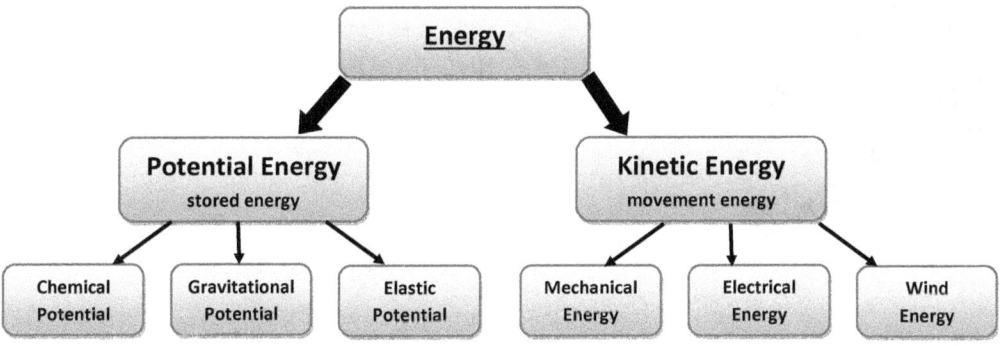

Chemical Potential Energy are energies stored in form of food, fossil fuel or other chemicals.

 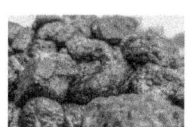 

Elastic Potential Energy is energy reside in spring, rubber or arrow bow.

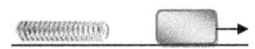

$$\text{EPE} = \frac{1}{2} k \cdot X^2$$

Elastic Energy = 1/2 Spring Constant x Extension²
(Joules)   (Newton/meter)   (meter)

Gravitational Potential Energy is the energy exerted on every object by the pull of the Earth.

$$GPE = m \times g \times h$$

Gravitational PE = mass x acc. due to gravity x height
(Joules)   (Kg)   (10 m/s²)   (meter)

Example 6.2

A ball of mass 0.5 kg is kicked vertically to the height of 15m. Calculate its gravitational potential energy at the highest point.

solution:

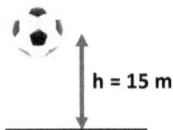

h = 15 m

GPE = m x g x h

GPE = 0.5 x 10 x 15 = <u>75 J</u>

Gravitational Potential Energy on the ball is 75 joules

Mechanical energy is the energy posses by a moving object.

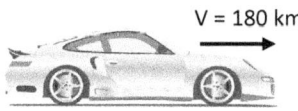

Velocity →

$$KE = \frac{1}{2} m \times v^2$$

Kinetic Energy = (mass x velocity²) ÷2
(Joules)   (Kg)   (m/s)

Example 6.2

A sports car of mass 1000 kg cruising with a velocity of 180 km/hr on a highway. Calculate its kinetic energy during the cruise.

solution:

V = 180 km/hr   Don't forget to convert velocity to m/s

KE = 0.5 x m x v²

KE = 0.5 x 1000 x (180 x 1000m ÷ 3600s)² = <u>1.25 x 10⁶ J</u>

Kinetic Energy of the car is 1.25 Megajoules

Law of conservation of energy

→ Energy is neither created nor destroyed

→ Energy is only being transformed from one to another

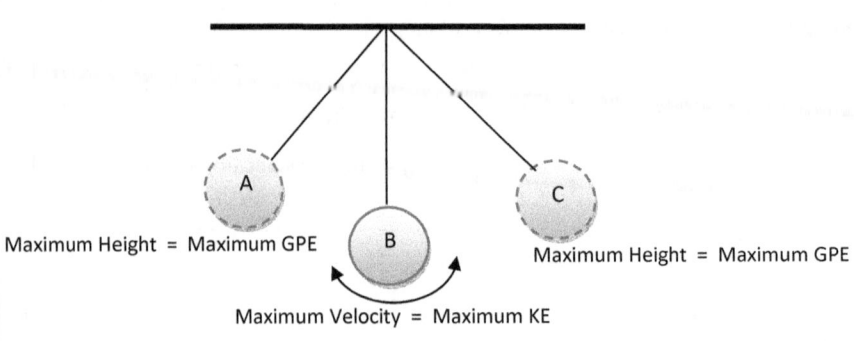

Notice that from the graph above total energy remains constant except for 'kinetic energy' and 'gravitational potential energy' that are interchanging with each other.

Example 6.3

A spring with a spring constant of 2 N/cm is compressed to 3 cm by a block of mass 0.4 kg, once the spring decompressed the mass is pushed. What is the velocity of the block ?

Solution:

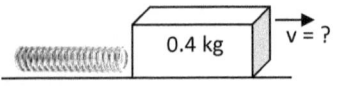

We know from law of conservation that
Elastic PE → Kinetic Energy

so we can say EPE = KE
$0.5 \, k \, x^2 = 0.5 \, m \, v^2$
$0.5 \, (2) \, (3)^2 = 0.5 \, (0.4) \, v^2$
v = <u>6.7 m/s</u>

Power

Power is rate of work done, or amount of energy used per unit time.

$$\text{Power (Watts)} = \frac{\text{Work}}{\text{time}} = \frac{\Delta \text{ Energy (Joules)}}{\Delta \text{ time (second)}}$$

Example 6.4

A crane lift a concrete block of mass of 250 kg to the height of 30 meters in 20 seconds. What is the rate of work done by the crane ?

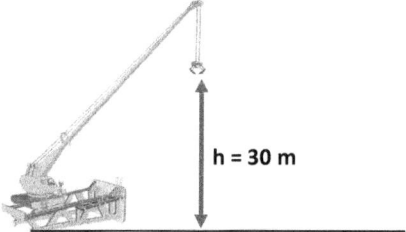

h = 30 m

Solution: We know that 'rate of work done' = Power

$$\text{Power} = \frac{\Delta \text{ GPE}}{\text{time}}$$

$$\text{Power} = \frac{m \times g \times h}{\text{time}}$$

$$\text{Power} = \frac{(250)(10)(30)}{20}$$

$$\text{Power} = \underline{3750 \text{ Watts}}$$

The crane uses the energy at the rate of 3750 joules per second.

Practice Chapter 6

Question 1-3

A security guard with a mass of 75 kg exerts a force of 350 N in pushing 2,000 kg car, the car moved a distance of 12 meters. Given that the car was pushed at constant speed in 5 seconds

1. The work done by the security guard in pushing the car is

    a)   840   J

    b)   4200   J

    c)   9000   J

    d)   12600   J

    e)   24000   J

2. The power developed in pushing the car is

    a)   840   J/s

    b)   920   J/s

    c)   1000   J/s

    d)   1260   J/s

    e)   2400   J/s

3. The frictional force that the car experience was

    a)   exactly 20,000 N

    b)   more than 20,000 N

    c)   between 350 to 750 N

    d)   less than 350 N

    e)   exactly 350 N

SAT Physics (Physics made simple)

4. A rock of mass 2 kg is dropped from the cliff 30 m high, ignoring air resistance, at what speed will the rock hit the ground ?

   a) 10   m/s

   b) 15.5 m/s

   c) 20   m/s

   d) 24.5 m/s

   e) 30   m/s

5. A 100 g marble lying on the table is pushed against the spring with a spring constant of 50 N/m, if the spring was compressed by 4 cm then how fast would the marble be pushed away ?

   a) 0.9 m/s

   b) 1.5 m/s

   c) 2.5 m/s

   d) 6.5 m/s

   e) 8.5 m/s

6. If the speed of the object is halved, what happen to its kinetic energy ?

   a) doubled

   b) halved

   c) quadrupled

   d) quartered

   e) tripled

7. According to law of conservation of energy, which of the following is true ?

   a) Energy can be converted and created

   b) Energy can be changed from one form to another

   c) Energy is a vector quantity

   d) Total energy of the system is equal to kinetic energy

   e) Work done is not equal to energy changed

8. How much work was done by load to extend the spring from 3 cm to 5 cm ?

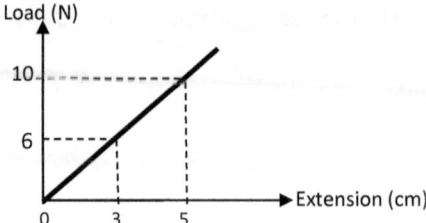

a) 0.16 J
b) 0.6 J
c) 6 J
d) 16 J
e) 60 J

Question 9 - 10

An elevator transport a load of 300 kg up 6 floors, each floor is 2.5 meter high. It's average speed is 30 m/s during the non-stop movement.

9. How much work was done by the elevator ?

a) 7.5 kJ
b) 20 kJ
c) 36 kJ
d) 40 kJ
e) 45 kJ

10. How much power was expended by the elevator ?

a) 32 kW
b) 60 kW
c) 80 kW
d) 90 kW
e) 135 kW

Answer: Chapter 6

Question	Answer	Explanation
1.	B	W = F x d W = 350 x 12 W = 4200 Joules
2.	A	Power = work/ time Power = 4200 / 5 Power = 840 watts
3.	E	Constant speed means that the resultant force or net force of zero Friction = F_{push} Friction = 350 N exactly
4.	D	GPE = KE mgh = 0.5mv² gh = 0.5 v² v² = 2 x g x h v² = 2 x 10 x 30 v = 24.5 m/s
5.	A	EPE = KE 0.5 k x² = 0.5 m v² (50)(0.04)² = 0.1 v² V = 0.9 m/s
6.	D	<table><tr><th></th><th>Mass</th><th>Velocity</th><th>KE</th></tr><tr><td>OLD→</td><td>m</td><td>v</td><td>0.5 m v²</td></tr><tr><td>NEW→</td><td>m (same)</td><td>0.5 v (halved)</td><td>0.5 m (0.5v)² [0.5 m v²]x 0.25</td></tr></table> The new KE is a quarter of old KE

SAT Physics (Physics made simple)

Question	Answer	Explanation
7.	B	Energy can only be transformed or ' changed from one form to another'
8.	A	Area under the graph will give us the work done in extending the spring Area = 1/2 x b x ($h_1 + h_2$) Area = 1/2 x 0.02m x (6 +10) Work = 0.16 Joules
9.	E	Work = ΔGPE Work = m x g x h Work = 300 x 10 x (2.5 x6) Work = 45000 Joules
10.	D	Power = Work ÷ time = F x **d÷t** = F x **v** =Weight x velocity Power = 3000 x 30 = 90,000 Watts Power = 90 kW

Momentum & Torque   7

Check-list:

- ✓ Momentum
- ✓ Torque (Turning effect of force)
- ✓ Angular momentum

Momentum

Momentum occurs during a collision or impact between bodies. We will examine only linear collision, or one-dimensional collision. Momentum is a vector quantity, a direction is always included with a positive or a negative sign.

$$\text{Momentum} = \text{Mass} \times \text{Velocity}$$
$$p = m\,v$$
$$(Kg \cdot m/s) \quad (kg) \; (m/s)$$

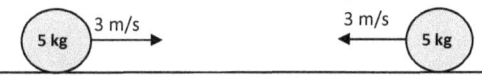

After the collision will both balls come to rest ?

To answer the question we need to understand 'Law of Conservation of Momentum'

Law of Conservation of Momentum

Total momentum = Total momentum
Before collision   After collision

$$\Sigma p_i = \Sigma p_f$$

SAT Physics (Physics made simple)

Example 7.1

What is the velocity (V_2) after the collision?

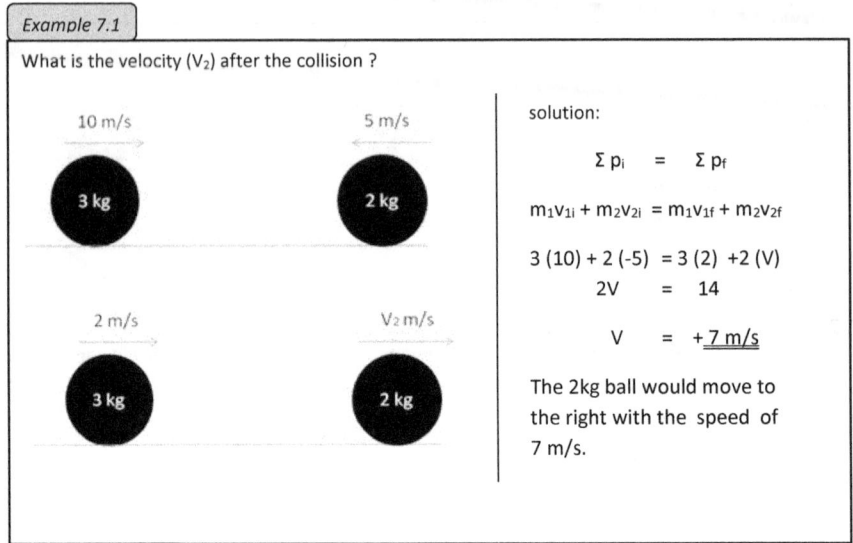

solution:

$$\Sigma p_i = \Sigma p_f$$
$$m_1v_{1i} + m_2v_{2i} = m_1v_{1f} + m_2v_{2f}$$
$$3(10) + 2(-5) = 3(2) + 2(V)$$
$$2V = 14$$
$$V = +\underline{7\ m/s}$$

The 2kg ball would move to the right with the speed of 7 m/s.

2 Kind of collisions

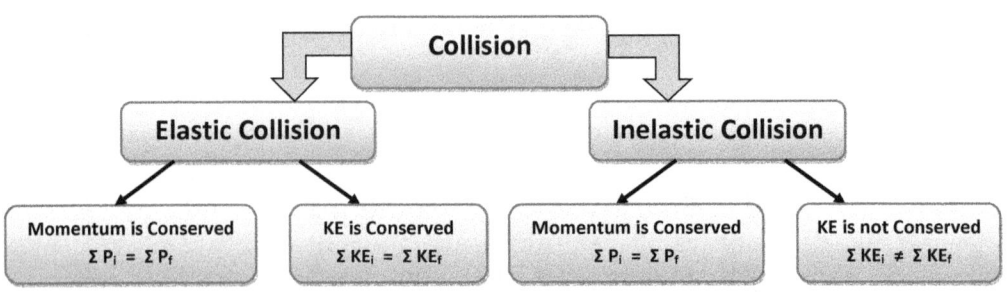

Impulse

Impulse is a change in momentum that occur in a complete collision. We can refer 'impulse' to area under the graph of Force VS Time graph.

$$\text{Impulse} = \text{Mass} \times \text{Change in Velocity}$$
$$\Delta p = m \, \Delta v$$
$$(Kg \cdot m/s) \quad (kg) \quad (m/s)$$

$$\text{Impulse} = \text{Force} \times \text{Change in Time}$$
$$\Delta p = F \, \Delta t$$
$$(N \cdot s) \quad (N) \quad (s)$$

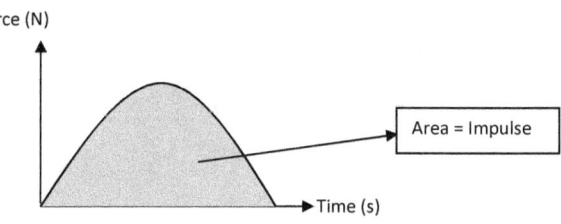

Area = Impulse

Example 7.2

According to example 7.1

a) what is the impulse on 2 kg ball ? b) what is the force of impact on 2 kg ball if the collision lasted for 0.2 seconds ?

Solution: a) $\Delta p = m \, \Delta v = m \, (V_f - V_i)$

$\Delta p = 2 \, (7 - (-5)) = 2 \, (14)$

$\Delta p = \underline{28 \text{ Kg} \cdot m/s}$

b) $\Delta p = F \, \Delta t$

$28 = F \, (0.2)$

$F = \underline{140 \text{ N}}$

Torque (turning effect of force)

Torque or moment is also known as turning effect of force that causes an object or a pivot to turn or rotate. Torque is a vector quantity, a direction here is denoted by clock-wise or anti clock-wise.

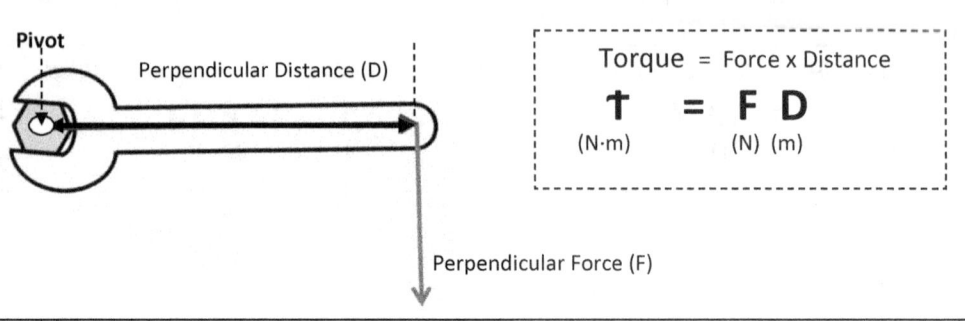

Torque = Force x Distance

$$\tau = F\ D$$

(N·m)　　(N) (m)

Example 7.3

Calculate the torque of the diagram

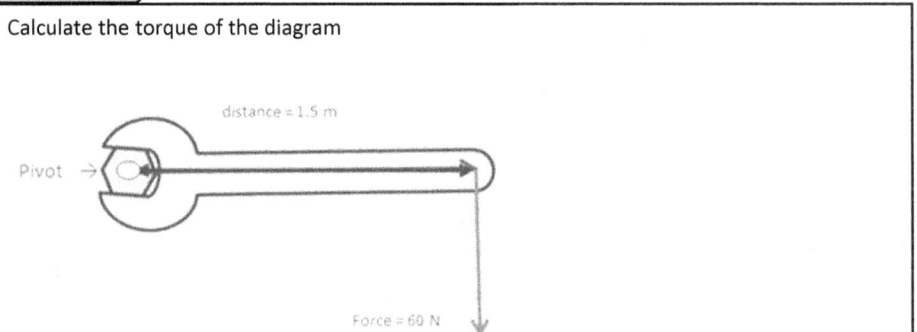

solution:

$\tau = F\ D = 60 \times 1.5 = \underline{90\ \text{N·m}}$

Clock-wise direction

Concept of torque are normally used in see-saw or fly-wheel. For see-saw to be balanced we have to

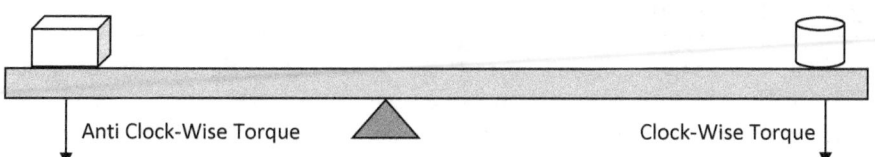

Anti Clock-Wise Torque   Clock-Wise Torque

In equilibrium (see-saw is balanced)

Total torque = Total torque
Anti clock-wise   Clock-wise

$$\Sigma \tau_{acw} = \Sigma \tau_{cw}$$

&

Total Upward Force = Total Downward Force

$$\Sigma F_{up} = \Sigma F_{down}$$

Example 7.4

What must be the weight of the girl W_2 ? What force is the exerted by the pivot to the see-saw? Given that the boy has a weight (W_1) of 900 N and the boy is 2.5 meter away from the pivot, while the girl is 3 meter away from the pivot.

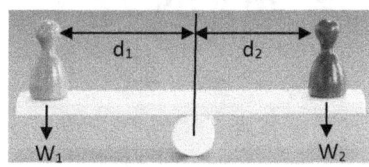

solution: a)

$$\Sigma \tau_i = \Sigma \tau_f$$

$$W_1 \times d_1 = W_2 \times d_2$$

$$900 \times 2.5 = W_2 \times 3$$

$$W_2 = \underline{750 \text{ N}}$$

b)

$$\Sigma F_{up} = \Sigma F_{down}$$

$$F_{pivot} = W_1 + W_2$$

$$F_{pivot} = 900 + 750$$

$$F_{pivot} = \underline{1650 \text{ N}}$$

Angular momentum

A system that has a net torque equals to zero will have zero rate of change in angular momentum, this means that angular momentum of a system is conserved. To understand angular momentum easier we should imagine objects moving in circular orbit such as planets. All the planets will have same rate of change in angular momentum.

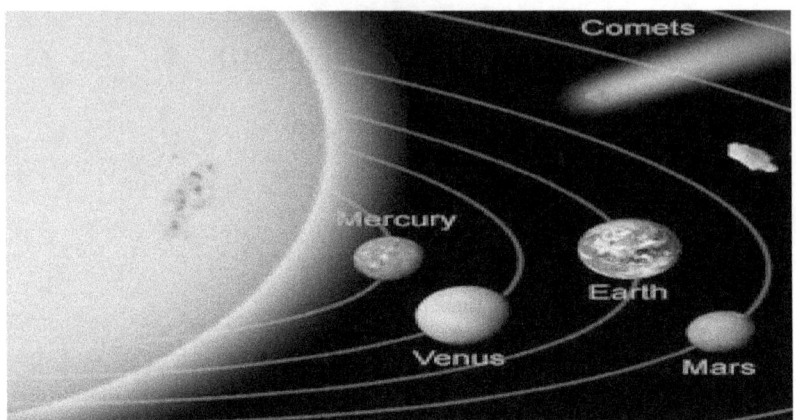

Angular momentum = mass x change in velocity x distance
$$\Delta L = m \times \Delta v \times r$$
$(kg \cdot m^2/s) \quad (kg) \quad (m/s) \quad (m)$

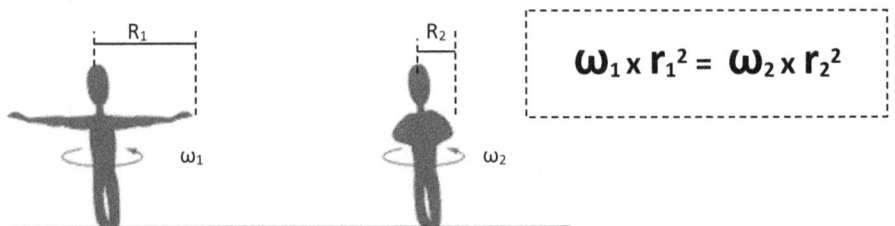

$$\omega_1 \times r_1^2 = \omega_2 \times r_2^2$$

** if you reduce your radius of rotation by half then you rate of rotation increased by factor of four

> **Example 7.5**
>
> Planet A has twice a mass of planet B, if the orbital velocity of both planet from the Sun is same then what is the ratio of orbital radius of planet A to planet B ?
>
> solution:
>
> $$\Delta L_{Planet\ A} = \Delta L_{Planet\ B}$$
> $$\Delta L_A = \Delta L_B$$
> $$m_A\ v_A\ r_A = m_B\ v_B\ r_B \quad \leftarrow \text{same velocity means cancel both v's}$$
> $$m_A\ r_A = m_B\ r_B \quad \leftarrow m_A = 2,\ m_B = 1$$
> $$2 \times r_A = 1 \times r_B$$
> $$r_A : r_B = 1 : 2$$
>
> So we can say that the ratio of radius of planet A to B is 1 to 2. It's easier to say that planet A has half the orbital radius compared to planet B or planet A is closer to the Sun than planet B.

Practice Chapter 7

Question 1-3

A truck of mass 5,000 kg moving with 30 m/s had a head-on collision with a car of mass 2,000 kg moving with 40 m/s. The car and the truck were in contact for 20 millisecond. After the collision the car reverse direction and move with 22.5 m/s.

1. What is the new velocity of the truck ?

   a) 5   m/s

   b) 10   m/s

   c) 15   m/s

   d) 20   m/s

   e) 25   m/s

2. What is the impulse experienced by the car ?

   a)   84000   N·s

   b)   92000   N·s

   c)   100000   N·s

   d)   125000   N·s

   e)   240000   N·s

3. The impact force that the car experience was

   a) 5.25×10^6 N

   b) 6.25×10^6 N

   c) 5.25×10^5 N

   d) 6.25×10^5 N

   e) 6.25×10^3 N

SAT Physics (Physics made simple)

4. A 10 g ball is thrown horizontally to the wall with a speed of 12m/s, it bounces back with the same speed. What is the magnitude of impulse on the ball ?

a)   24   N·s

b)   2.4   N·s

c)   1.2   N·s

d)   0.24 N·s

e)   0.12 N·s

5. Which of the following statement is true about elastic collision ?

a)   KE is conserved

b)   Only PE is lost during the collision

c)   PE and KE is not conserved

d)   All energy is loss during the collision

e)   none of the above

6. If the radius of rotation of a planet is tripled, what happen to its rate of rotation ?

a)   tripled

b)   reduced by 1/3 times

c)   increased by 9 times

d)   reduced by 1/9 times

e)   no change

7. A 50 kg boy sits on a seesaw 0.5 m away from the pivot, how far on the other end should a 75 kg girl sits from the pivot ?

a)   0.5   m

b)   0.45 m

c)   0.4   m

d)   0.33 m

e)   0.25 m

8. How much impulse is deliver during the collision represented by graph below ?

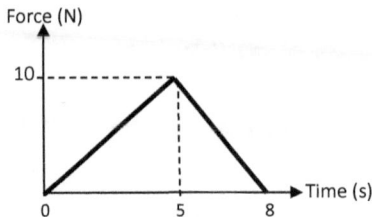

a) 20   N·s

b) 30   N·s

c) 40   N·s

d) 60   N·s

e) 80   N·s

Question 9 - 10

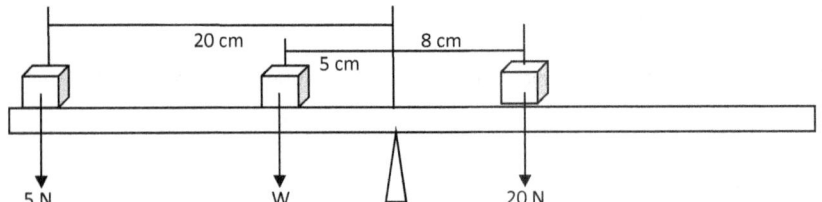

9. If the seesaw above is balanced then what is the value of 'W' ?

a) 7.5   N   d) 22   N

b) 12   N   e) 25   N

c) 18   N

Scan me

10. What is the force acting on the pivot ?

a) 25   N   d) 37   N

b) 30   N   e) 50   N

c) 32.5 N

Answer: Chapter 7

Question	Answer	Explanation
1.	A	$\Sigma P_i = \Sigma P_f$ $(5000)(30) + (2000)(-40) = (5000)V + (2000)(22.5)$ $V = 5$ m/s
2.	D	$\Delta P = m \Delta v = m(v_f - v_i)$ $\Delta P = (2000)(22.5 - (-40))$ $\Delta P = 125,000$ N·s
3.	B	$\Delta P = F \Delta t$ $125000 = F(20 \times 10^{-3})$ $F = 6.25 \times 10^6$ N
4.	D	$\Delta P = m \Delta v = m(v_f - v_i)$ $\Delta P = (0.01)(-12 - 12)$ $\Delta P = 0.24$ N·s
5.	A	Elastic collision means that momentum and kinetic energy are conserved always. There is no loss of energy during the collision.
6.	D	Rate of rotation means → $\omega_1 \times r_1^2 = \omega_2 \times r_2^2$ |	Rate of rotation (ω)	Radius (r)	ω x r² | |---|---|---|---| | OLD→ | 1 | 1 | 1 | | NEW→ | ω_2 = ??? | 3 (tripled) | 1 (conserved) | $\omega_1 \times r_1^2 = \omega_2 \times r_2^2$ $1 = \omega_2 \times (3)^2$ $\omega_2 = 1/9$

SAT Physics (Physics made simple)

Question	Answer	Explanation
7.	D	$\Sigma \tau_{acw} = \Sigma \tau_{cw}$ $W_1 \times d_1 = W_2 \times d_2$ $(50)(0.5) = 75(d)$ $d = 0.33$ m
8.	C	Area under the graph will give us the impulse Area $= 1/2 \times b \times h$ Area $= 1/2 \times 8 \times 10$ Impulse $= 40$ N·s
9.	B	$\Sigma \tau_{acw} = \Sigma \tau_{cw}$ $(5)(20) + w(5) = (20)(8)$ $w = 12$ N
10.	D	$\Sigma F_{up} = \Sigma F_{down}$ $F_{pivot} = 5 + 12 + 20$ $F_{pivot} = 37$ N

Gas & Pressure 8

Check-list:

- ✓ Pressure
- ✓ Kinetic Theory
- ✓ Gas Law

Pressure

Pressure is a measure of force applied per unit area. Pressure is also called 'stress' and its unit is Pascal. Pressure is directly proportional to force applied, and pressure is inversely proportional to area.

$$\text{Pressure} = \text{Force} \div \text{Area}$$

$$P_{(Pa)} = \frac{F_{(N)}}{A_{(m^2)}}$$

Example 8.1

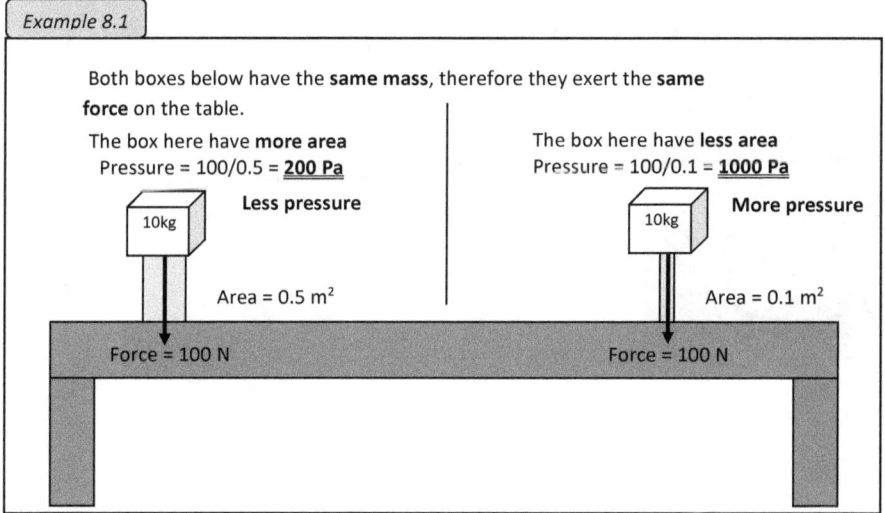

Both boxes below have the **same mass**, therefore they exert the **same force** on the table.

The box here have **more area**
Pressure = 100/0.5 = **200 Pa**
Less pressure
10kg
Area = 0.5 m²
Force = 100 N

The box here have **less area**
Pressure = 100/0.1 = **1000 Pa**
More pressure
10kg
Area = 0.1 m²
Force = 100 N

SAT Physics (Physics made simple)

Pressure in liquid

Concept of pressure in liquid:

    i. Pressure acts in all direction
    ii. Pressure increases with depth
    iii. Pressure is transmitted throughout liquid

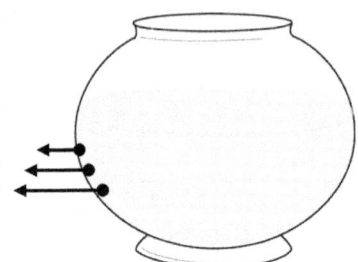

Pressure = depth x density x acc. due to gravity

$$P = h \rho g$$

(Pa)   (m)   (kg/m³)   (10 m/s²)

Example 8.2

What is pressure at the bottom of the aquarium of depth 15 meter ? Given that density of salt water is 1020 kg/m³

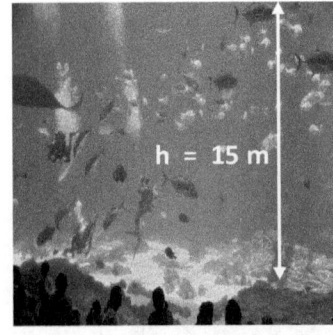

h = 15 m , ρ = 1020 kg/m³ , g = 10 m/s²

$$P = h \rho g$$

P = 15 x 1020 x 10

P = <u>153,000 Pascal</u>

Total pressure at the bottom is 153 KPa

3 states of matter

Particles	Molecular Structure/bond	Definite Shape	Definite Volume
Solid	Particles are **closely packed.** Molecular bond are **very strong.**	YES	YES
Liquid	Particles are **loosely packed.** Molecular bond are **not so strong.**	NO	YES
Gas	Particles are **far apart.** Molecular bond are **very weak.**	NO	NO

Kinetic Theory of Gas

Concept of kinetic theory of gas can be summarize as follow :

i. Gases are made up of molecules

ii. Molecules are in constant random motion

iii. The movement of molecules is governed by Newton's Laws

iv. Molecular collisions are perfectly elastic

Total kinetic and potential energies of all the atoms or molecules in a material is known as **internal energy**. The hotter the material is, the more internal energy it has, the faster particles move.

Temperature is a measure of degree of hotness or coldness, it is also a measure of average kinetic energy per molecule of a substance. **Heat** is the energy that flow between bodies of different temperature, heat always flow from hot to cold.

Absolute zero

Absolute zero is the lowest possible temperature where nearly no heat energy remains in particles. Absolute zero is the point at which the particles of have minimal vibration, zero-point energy-induced particle motion. As we increase the temperature, the kinetic energies of the particles increases.

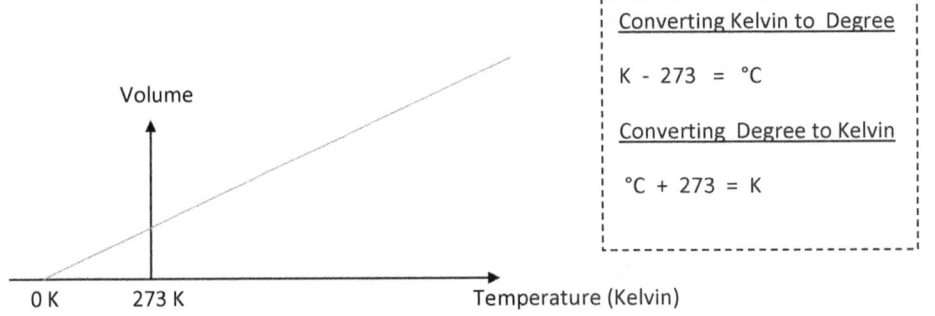

Converting Kelvin to Degree

$K - 273 = °C$

Converting Degree to Kelvin

$°C + 273 = K$

*We can say that absolute temperature is directly proportional to kinetic energy of the particles.

Mercury Barometer

Mercury barometer is used to measure atmospheric pressure by using the height of mercury.

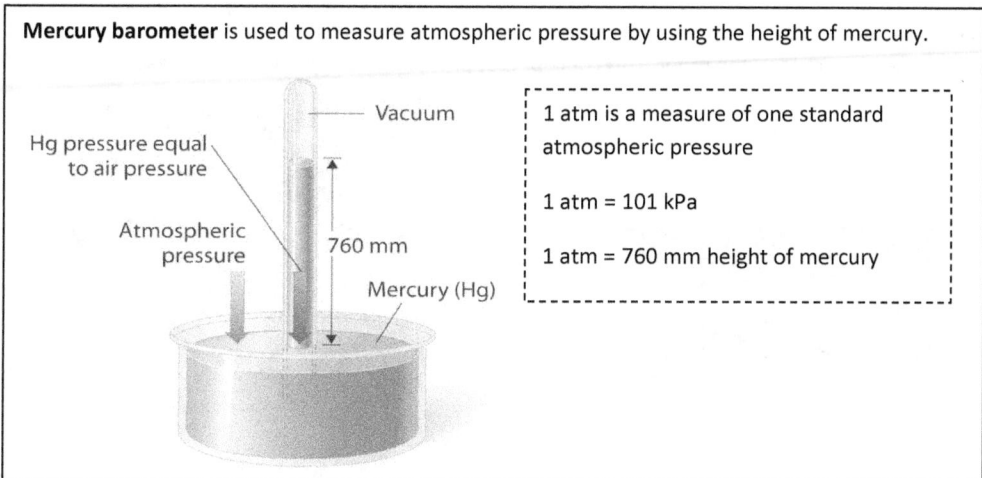

1 atm is a measure of one standard atmospheric pressure

1 atm = 101 kPa

1 atm = 760 mm height of mercury

Thermometer

Thermometer uses expansion of liquid(Alcohol or Mercury) to measure the temperature.

There are two fixed point on a thermometer:

i) **Upper fixed point** is measured at 100°c by putting the thermometer in the boiling water

ii) **Lower fixed point** is measured at 0°c by putting the thermometer in the melting ice

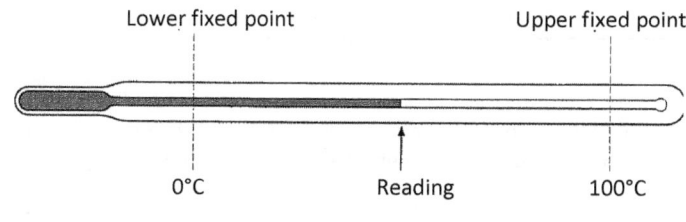

Expansion of Solid

Here is an example of a bimetallic strip, heat are applied to both the metals at the same time. One metal expand more than the other. This concept is applied on building roads or even gap in rail way.

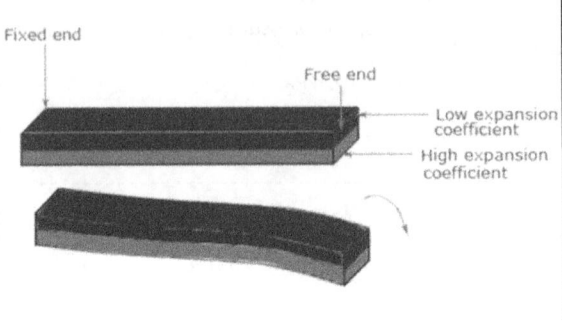

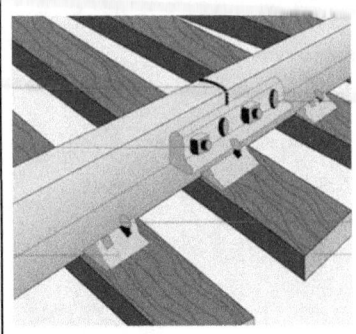

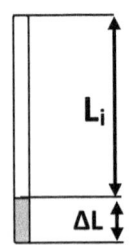

Change in Length = coefficient of expansion x length x change in temperature

$$\Delta L = \alpha \; L_i \; \Delta T$$
$$(m) \quad (°C^{-1}) \; (m) \quad (°C)$$

Change in volume = coefficient of expansion x length x change in temperature

$$\Delta V = \beta \; V_i \; \Delta T$$
$$(m^3) \quad (°C^{-1}) \; (m^3) \quad (°C)$$

Example 8.3

An aluminum box with a dimension of 0.2 m by 0.1 m by 0.3m has a coefficient of volume expansion of 69×10^{-6} /°C. If it was heated from the temperature of 20°C to 90°C then what is the change in volume ?

solution:

$V_i = 0.2 \times 0.1 \times 0.3 = 0.006 \; m^3$, $\Delta T = 90 - 20 = 70 \; °C$, $\beta = 69 \times 10^{-6}$

$\Delta V = \beta \; V_i \; \Delta T = 69 \times 10^{-6} \times (0.006) \times (70)$

$\Delta V = \underline{0.000029 \; m \; or \; 29 \; \mu m}$

Gas Law

Whenever we are using gas law or dealing with gases we must fixed its mass. The three main properties of gas that we are concerned with are its volume, temperature and pressure.

*Remember while using 'Gas Law' the temperature must always be converted to Kelvin scale.

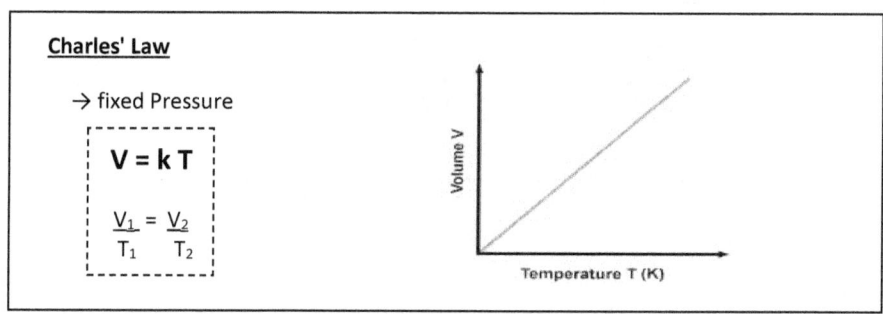

Charles' Law

→ fixed Pressure

$$V = kT$$

$$\frac{V_1}{T_1} = \frac{V_2}{T_2}$$

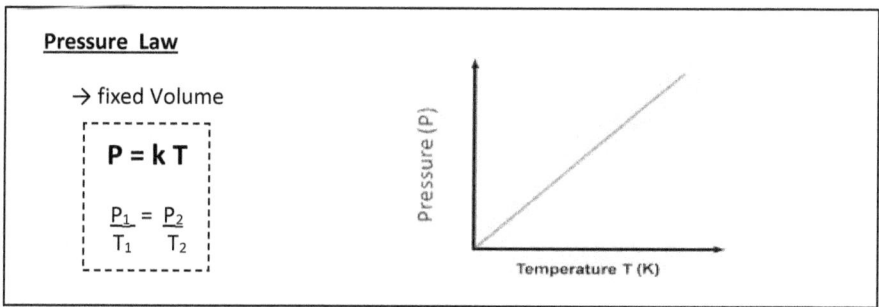

Pressure Law

→ fixed Volume

$$P = kT$$

$$\frac{P_1}{T_1} = \frac{P_2}{T_2}$$

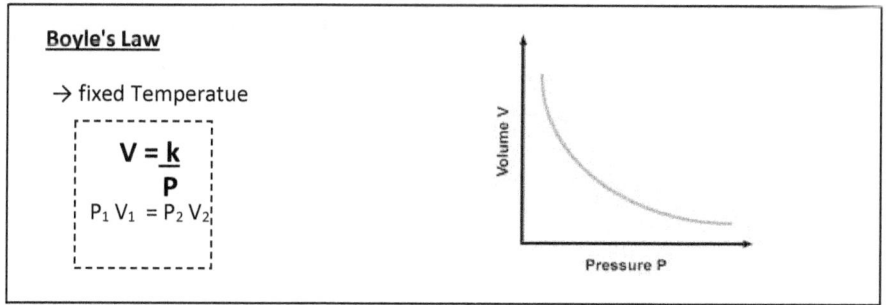

Boyle's Law

→ fixed Temperatue

$$V = \frac{k}{P}$$

$$P_1 V_1 = P_2 V_2$$

Ideal Gas Law

The absolute pressure of an ideal gas is directly proportional to Kelvin temperature and number of moles of the gas and is inversely proportional to the volume of the gas.

Combining all the laws we get:

$$\frac{P_1 V_1}{T_1} = \frac{P_2 V_2}{T_2}$$

Example 8.4

A box contain air of volume 5 cm³ at 1 atm is compressed to 3 cm³. Given that temperature and mass remain constant, what is the new pressure in kPa of air in this box ?

solution:

$P_1 V_1 = P_2 V_2$

1 atm = 101 kPa → 101 kPa (5 cm³) = P (3 cm³)

P = <u>168.33 kPa</u>

So the pressure of air in the box will be raise to 168.33 KiloPascal.

Practice Chapter 8

1. A submarine travelling from Russia to Alaska is at the depth of 400 m, at this depth what is the pressure on the submarine ?(Given that density of salt water is 1020 kg/m³)

    a) 4080   Pa
    b) 408   kPa
    c) 40.8   kPa
    d) 4.08   MPa
    e) 0.408   MPa

2. The windows of the submarine each has an area of 400 cm², what is the force exerted on a window of the submarine in question 1 ?

    a)   163.2   kN
    b)   16.32   kN
    c)   1.632   kN
    d)   0.163   kN
    e)   0.163   MN

3. Which of the following is NOT TRUE about 'kinetic theory of gas' ?

    a)   Air particles and gas are made up of molecules
    b)   Gas molecules are in constant random motion
    c)   The movement of molecules is governed by Newton's Laws
    d)   Molecular collisions are perfectly elastic
    e)   Total kinetic and potential energies of a material are equals

4. Convert 20°C to Kelvin

   a) 253 K

   b) 283 K

   c) 293 K

   d) 303 K

   e) 313 K

5. A concrete slab with the length of 30 cm has a coefficient of linear expansion of 12×10^{-6} /°C. If it was heated from the temperature of 25°C to 55°C then what is the increase in length?

   a) 0.36   cm

   b) 0.108 cm

   c) 0.092 cm

   d) 0.024 cm

   e) 0.011 cm

Scan me

6. An air particles in a box it heated so that the pressure of air inside is tripled. What happen to the volume ?

   a) tripled

   b) reduced by 1/3 times

   c) increased by 9 times

   d) reduced by 1/9 times

   e) no change

7. A gas in a cylinder at temperature of 27°C has a volume of 20 cm³, what is the temperature of the gas if the volume is increased to 30 cm³ ?

   a) 40.5   °C

   b) 54   °C

   c) 177   °C

   d) 300   °C

   e) 450   °C

8. 1.5 atm is equal to approximately

   a) 152   Pa

   b) 152   kPa

   c) 76   cm of mercury height in barometer

   d) 104   cm of mercury height in barometer

   e) 103   N/m²

9. Which of the following method is used to find lower fixed point in a thermometer ?

   a)   put thermometer in boiling water

   b)   put thermometer in melting ice

   c)   put thermometer in boiling mercury

   d)   put thermometer in melting mercury

   e)   put thermometer in boiling alcohol

10. Which quantity is represented by the product of pressure and change in volume of a piston pushed ?

    a)   Temperature

    b)   Mass

    c)   Coefficient of volume expansion

    d)   Power

    e)   Work done

Answer: Chapter 8

Question	Answer	Explanation
1.	D	$P = h \rho g$ $P = 400 \times 1020 \times 10$ $P = 4{,}080{,}000$ Pa $= 4.08$ MPa
2.	A	$P = F \div A$ $F = P \times A$ $F = (4.08 \times 10^6) \times [\,400 \times (10^{-2})^2\,]$ $F = 163{,}200$ N $= 163.2$ kN
3.	E	All of the choices from A to D are true (Hence they are listed in the fifth page of this chapter) All PE ≠ All KE The only wrong choice is E
4.	C	To convert Degree to Kelvin we add 273 20 + 273 = 293 K
5.	E	$\Delta L = \alpha\, L_i\, \Delta T$, $L_i = 30$ cm , $\Delta T = 55 - 25 = 30\,°C$, $\alpha = 12 \times 10^{-6}$ $\Delta L = (12 \times 10^{-6})(30)(30)$ $\Delta L = 0.0108$ cm
6.	E	In this question there are only relationship between volume and temperature, only two can change at a time (unless it is said to be an ideal condition). So only change here is between volume and temperature but pressure is fixed. No change in pressure occur here.

Question	Answer	Explanation
7.	C	We use Charles' Law here (change temperature to Kelvin) $\dfrac{V_1}{T_1} = \dfrac{V_2}{T_2}$ $\quad T_1 = 27 + 273 = 300$ K $\dfrac{20}{300} = \dfrac{30}{T_2}$ $T_2 = 450$ K $T_2 = 450 - 273 = 177\ °C$
8.	B	1 atm = 101 kPa 1.5 atm = 1.5 x 101 kPa 1.5 atm = 151.5 or about 152 kPa
9.	B	To find the lower fixed point of thermometer we always look for 0°C, which is the point where ice melt
10.	E	Pressure x Change in volume = P x A x Δd $\quad\quad\quad\quad\quad$ (P X A = F) → (F) x Δd = Work

SAT Physics (Physics made simple)

Heat　　　　　　　　　　　　　　　　　　　　　　　　　　　9

Check-list:

- ✓ Heat capacity
- ✓ 3 Ways heat travel
- ✓ Law of thermodynamic

Heat Capacity

Specific heat capacity is amount of energy needed to raised 1°C of 1 kg substance. For example if we have a cup of tea that contain 1 kg of tea at temperature of 25 °C and we want to raise its temperature to 26 °C we would need to give the energy to it

Heat = mass x specific heat capacity x change in temperature

$$Q = m \cdot c \cdot \Delta T$$

(Joules)　(kg)　(J/kg·°C)　(°C)

Substance	Specific heat capacity (J/kg·°C)
Water	4200
Ethanol	2410
Copper	381
Aluminum	878
Iron	438
Lead	126
Ice	2110

Example 9.1

How much energy is needed to heat 12 kg of water at 31 °C to 41°C ?

solution:　m = 12 kg, ΔT = 41 - 31 = 10 °C, C_{water} = 4,200 J/kg·°C

$$Q = m \times c \times \Delta T = 12 \times 4200 \times 10 = \underline{504{,}000 \text{ Joules}}$$

The amount of heat required to raise the temperature by 10°C is 504,000 Joules

Changing Phases

Energy of fusion is required to change the phase of the object from solid to liquid, while there should be no change in temperature of the substance.

Solid → Liquid

Heat of fusion = mass x latent heat of fusion

$$Q_f = m \cdot H_f$$
(Joules) (kg) (J/kg)

Ice at 0°C

+ Q_f =

Water at 0°C

Energy of vaporization is required to change the phase of the object from liquid to gas, while there should be no change in temperature of the substance.

Liquid → Gas

Heat of vaporization = mass x latent of vaporization

$$Q_v = m \cdot H_v$$
(Joules) (kg) (J/kg)

+ Q_v =

Substance	Latent Heat of Fusion (kJ/kg)	Melting point (°C)	Latent Heat of Vaporization (kJ/kg)	Boiling point (°C)
Water	334 (ice)	0	2260 (water)	100
Alcohol	108	-114	855	78.3
Ammonia	339	-75	1369	-33.3
Lead	24.5	327.5	871	1750

SAT Physics (Physics made simple)

Energy Transfer

```
┌─────────────────────┐     ┌─────────────────────────┐
│  Energy Loss        │  =  │  Energy Gain            │
│  by one substance   │     │  by another substance   │
└─────────────────────┘     └─────────────────────────┘
```

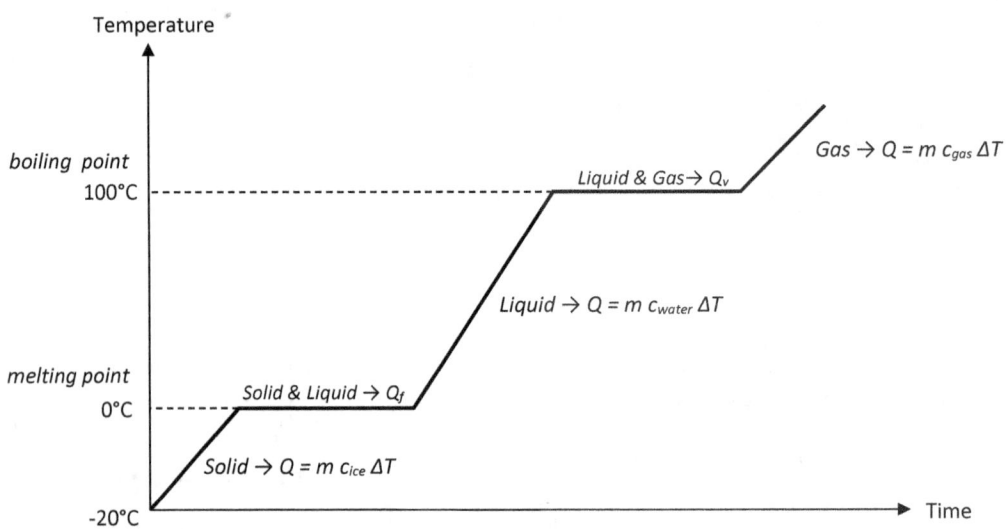

Example 9.2

How much energy is required to melt 5 kg of ice completely from -10°C to 55°C ?

solution: Total Energy = $Q_1 + Q_2 + Q_3$

Q_1 = Energy required to raise ice from -10 to 0°C

m = 5 kg, ΔT = 0 - (-10) = 10 °C, C_{ice} = 2,110 J/kg·°C

$Q_1 = m \times c \times \Delta T = 5 \times 2110 \times 10$ = **105,500 J**

Q_3 = Energy required to raise water from 0°C to 55 °C

, ΔT = 55 - (0) = 55 °C, C_{water} = 4,200 J/kg·°C

$Q_3 = m \times c \times \Delta T = 5 \times 4,200 \times 55$ = **1,155,000 J**

Q_2 = Energy required to melt ice

m = 5 kg, H_f = 334 kJ/kg·°C

$Q_2 = m \times H_f = 5 \times 334 \times 10^3$ = **1,670,000 J**

Total Energy = $Q_1 + Q_2 + Q_3$

Total Energy = 105.5k + 1.67M + 1.155M

Total Energy = **2,930,500 J**

3 Ways heat travel

Conductor allows heat and electrons to flow pass thru easily, all metal are good conductors.

Insulator do not allow heat and electrons to flow pass, good examples are rubber, cloth and plastic.

Heat always travel from hot to cold, the method of heat transfer are:

1. **Conduction** is a method of heat transfer that occurs only in solid. If we heat a metal its molecules gain kinetic energy then they all start to vibrate and collide elastically with each other.

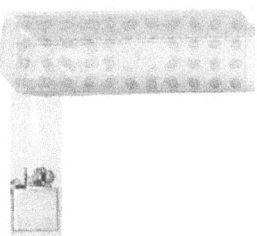

2. **Convection** is a method of heat transfer that occurs only in liquid and gas. If a liquid is heated the hot liquid become less dense and rises up, while cold liquid become more dense and falls.

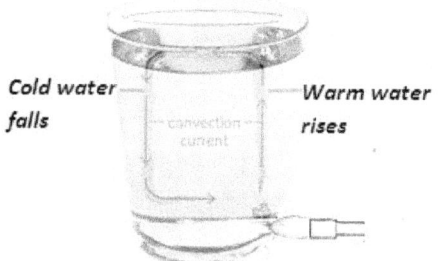

3. **Radiation** is a method of heat transfer in form of electromagnetic wave, infrared. This method require no medium, it can travel thru vacuum.

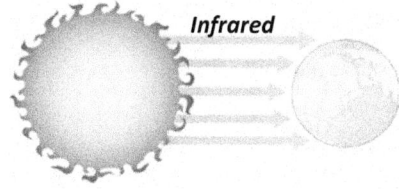

Law of thermodynamic

First law of thermodynamic state to total energy in a system is conserved. In other words the heat(Q) supplied to the system is equal to the sum of increase in internal energy(Δu) and work done(w) by the system.

$$Q = \Delta U + W$$

Work done on the system → W = +   Work done by the system → W = -

Heat added to the system → Q = +   Heat removed from the system → Q = -

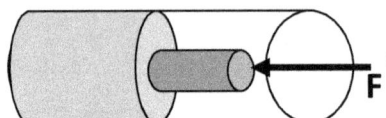 Work is done to the system since volume is decreasing

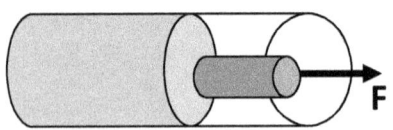 Work is done by the system since volume is increasing

Isothermal is a process where no change in temperature($\Delta T = 0$), so there is no change in internal energy ($\Delta U = 0 \rightarrow Q = W$)

Isochoric is a process where there is no change in the volume ($\Delta V = 0$), so there is no work done (W = 0 → Q = ΔU)

Isobaric is a process where there is no change in pressure ($\Delta P = 0$)

Adiabatic is a process where no heat is added or removed from an isolated system

Second law of thermodynamic state that :

i) Heat cannot flow from cold to hot

ii) Entropy of a system is always increasing until the system reached equilibrium

iii) At equilibrium point, entropy remains constant

Thermodynamic cycle can be represented by **Carnot cycle** below

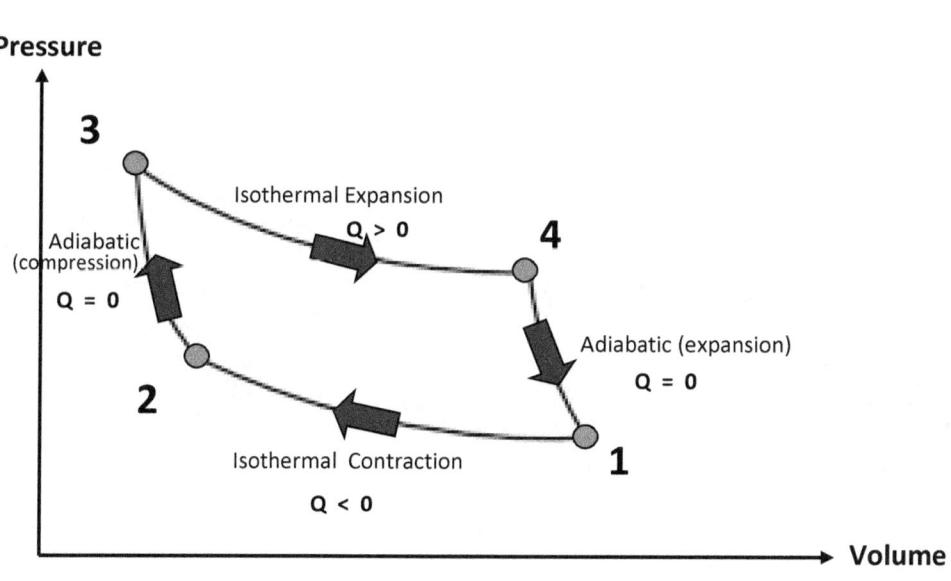

Example 8.4

An isolated system gives off 250 joules of heat, while the internal energy of the system is decreased by 400 joules. How much work was done by the system ?

solution:

$$Q = \Delta U + W$$
$$-250 = -400 + W$$
$$W = \underline{150 \text{ Joules}}$$

We can conclude that the system does 150 Joules of work.

Thermal conductivity

Dull black object are generally a good absorber and a good emitter of heat radiation but they are poor reflector of heat radiation.

Shinny white object are generally a poor absorber and a poor emitter of heat radiation but they are good reflector of heat radiation.

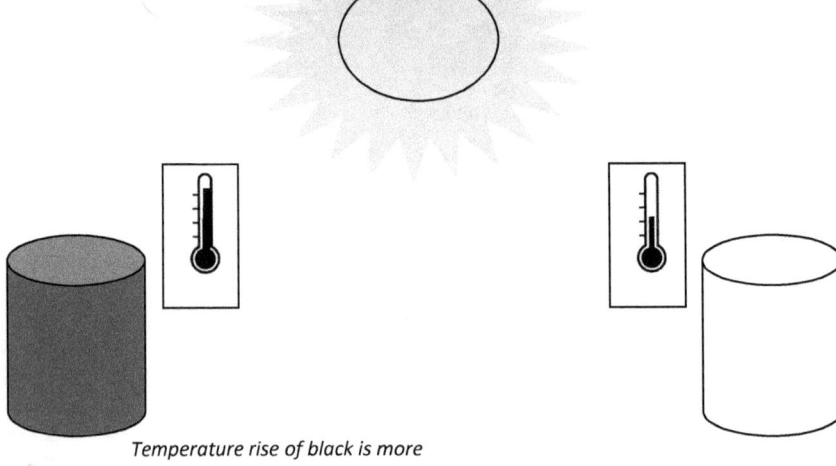
Temperature rise of black is more

Practice Chapter 9

1. A tub contain 50 kilogram of water at 10°C is heated by 60 kW heater for 5 minutes. How much energy is delivered to the water ?(Assuming no heat loss)

    a) 180   kJ

    b) 300   kJ

    c) 18   MJ

    d) 30   MJ

    e) 300   MJ

2. According to question 1, what is the final temperature of water after heat is supplied ?

    a) 60.7   °C

    b) 85.7   °C

    c) 90.7   °C

    d) 95.7   °C

    e) 102.7 °C

3. Ten-thousand joules of heat is added to 4 kilogram of ice at 0°C, what will be the final temperature after the heat is added?

    a) 0   °C

    b) 0.8   °C

    c) 10.7   °C

    d) 20.8   °C

    e) 22.7   °C

4. Which of the following is NOT TRUE about 'law of thermodynamic' ?

   a) Heat can flow from cold to hot

   b) Entropy remains constant at equilibrium point

   c) Entropy of a system is always increasing

   d) Heat cannot flow from hot to cold

   e) Reversing heat flow require external energy

5. How much energy is needed approximately to heat 200g of ethanol at 23 °C to 60°C ?

   a) 310   kJ

   b) 180   kJ

   c) 155   kJ

   d) 31   kJ

   e) 18   kJ

6. A group of students camping lit a camp fire and sits near it receive the heat chiefly by which method of heat transfer ?

   a) conduction and convection

   b) conduction and radiation

   c) conduction only

   d) convection only

   e) radiation only

7. If a can of water is left under the sun for one hour, which colored can would have the greatest rise in temperature ?

   a) dull black

   b) dull white

   c) shinny black

   d) shinny white

   e) matte green

8. A gas in a cylinder has a volume of 30 cm³ undergoes isobaric compression at 1 atm to volume of 50 cm³, how much work is done on the gas?

   a) 1.50   J

   b) 2.02   J

   c) 3.03   J

   d) 4.03   J

   e) 4.28   J

9. During isothermal expansion, 300 J of heat is supplied to the system. How much work is done on the system?

   a) 0   J

   b) 150   J

   c) 300   J

   d) 450   J

   e) 600   J

10. First law of thermodynamic is analogous to which law below?

    a) Newton's First law of motion

    b) Newton's Second law of motion

    c) Law of conservation of moment

    d) Law of conservation of momentum

    e) Law of conservation of energy

Scan me

11. Which quantity is represented by the quotient of heat and change in temperature?

    a) Thermal resistivity

    b) Thermal capacity

    c) Conducting coefficient

    d) Power

    e) Work done

SAT Physics (Physics made simple)

Answer: Chapter 9

Question	Answer	Explanation
1.	C	Power $= Q \div T$ Q (heat) $= P \times T$ $Q = (60 \times 10^3) \times (5 \times 60)$ $Q = 18,000,000$ J $= 18$ MJ
2.	D	$Q = m \times c \times \Delta T$ $18 \times 10^6 = 25 \times (4200) \times \Delta T$ $\Delta T = 85.7$ $T_f - T_i = 85.7$ $T_f = 85.7 + 10$ $T_f = 95.7$ °C
3.	A	Read the question carefully we will see that it is still "ice" We need to apply latent heat of fusion first $Q_f = m \times H_f$ $Q_f = 4 \times 334,000$ $Q_f = 1,336,000$ Joules is need to melt ice We supply only 10,000 J, which is less than Q_f So there is no change in temperature.
4.	D	Heat always flow from high temperature (hot) to lower temperature (colder). Choice D contradict the second law directly.
5.	E	$Q = m \times c \times \Delta T$ $Q = 0.2 \times 2410 \times (60 - 23)$ $Q = 17834$ J $Q = 18$ kJ
6.	E	In this question heat can only be transfer by radiation in form of infrared rays, since the students are sitting in front of the fire.

SAT Physics (Physics made simple)

Question	Answer	Explanation
7.	A	Dull black are the best absorber of heat radiation (infrared)

8.	B	Isobaric means $\Delta P = 0$ [1 atm = 101 kPa]
		 Work done = $P \Delta V$
		             = 101kPa x (50 - 30)x10^{-6}
		             = 2.02 Joules

9.	C	During isothermal expansion $\Delta U \rightarrow Q = W$ Q = 300 J W = $\underline{300\ Joules}$
10.	E	First law of thermodynamic $Q = \Delta U + W$ This is the law of conservation of energy
11.	B	Thermal conductivity measures how much energy is needed per degree increment. Thermal conductivity = $\dfrac{Q}{\Delta T}$

SAT Physics (Physics made simple)

Wave I   10

Check-list:

- ✓ Type of waves
- ✓ Wave properties
- ✓ Electromagnetic waves
- ✓ Mirror and Lens

Type of waves

Wave is a method of transfering energy from one point to another point.

2 Kinds of wave are:

1. Transverse wave

→ wave that oscillate or propagate perpendicular to the direction of motion

→ wave travels up and down

→ example: water waves, radio wave, infrared, visible light, UV, X-rays, gamma rays

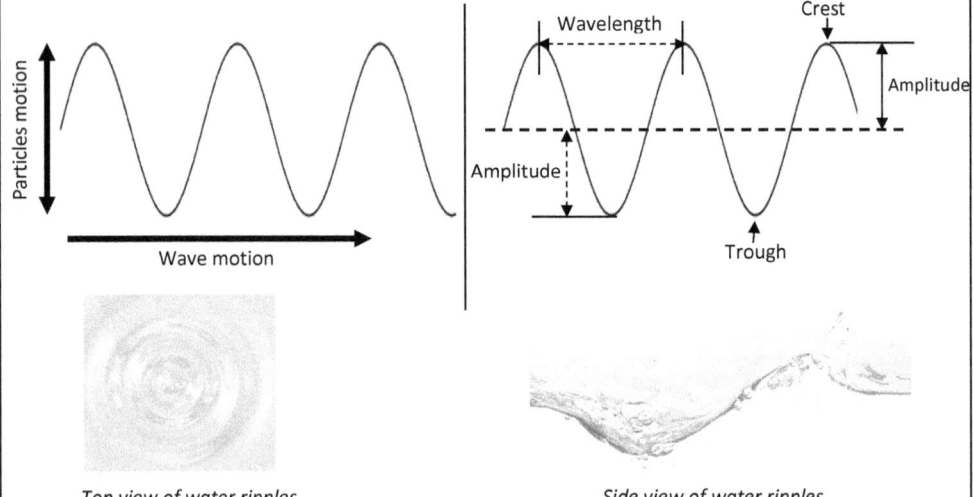

Top view of water ripples   *Side view of water ripples*

2. Longitudinal wave

→ wave that oscillate or propagate parallel to the direction of motion

→ wave travels back and forth

→ example: sound waves and seismic wave (p-wave and s-wave)

→ sound waves are vibration of air particles, composed of rarefaction (less pressure) and compression (high pressure)

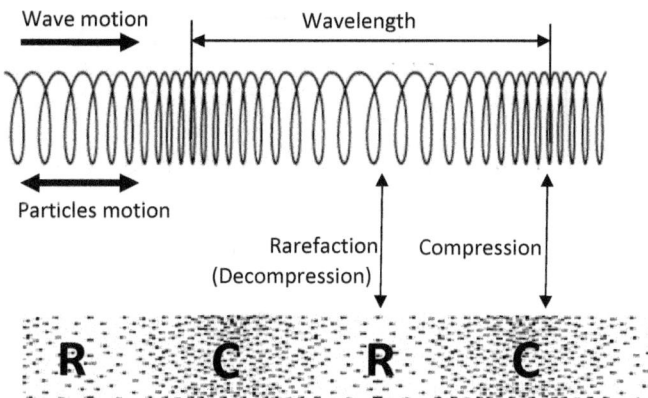

Example of sound wave in the air

Wave properties

Frequency and Period of wave

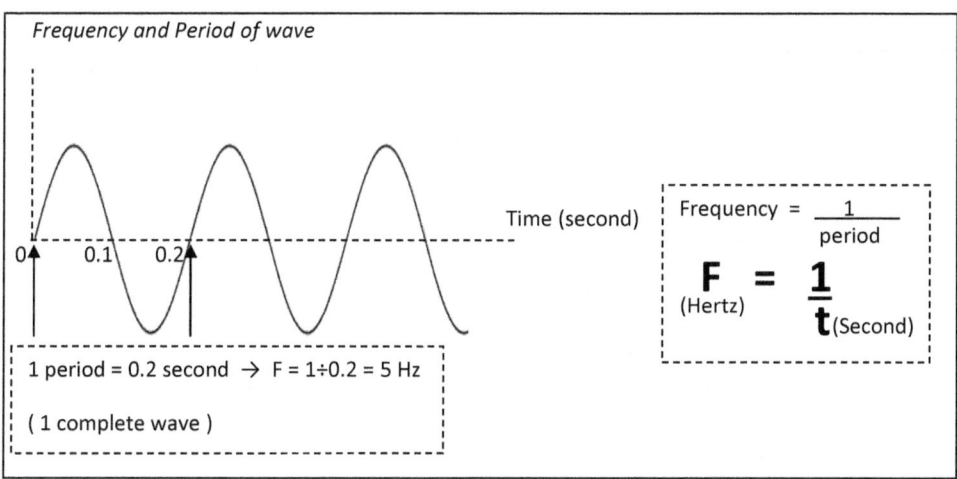

1 period = 0.2 second → F = 1÷0.2 = 5 Hz

(1 complete wave)

Frequency = $\dfrac{1}{period}$

$F_{(Hertz)} = \dfrac{1}{t_{(Second)}}$

SAT Physics (Physics made simple)

> Wave properties

Wave speed = frequency × wavelength
$$V = F\lambda$$
(m/s) (Hz) (m)

Pitch is affected by number of wave per second or frequency

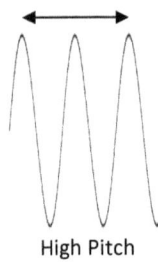
High Pitch

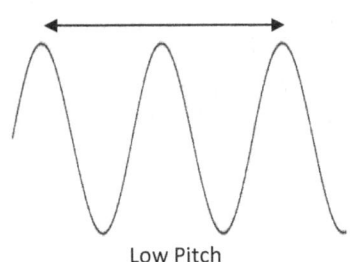
Low Pitch

** Human audible range of frequency are between 20 Hz to 20,000 Hz

Loudness is affected by amplitude of wave or energy transmitted

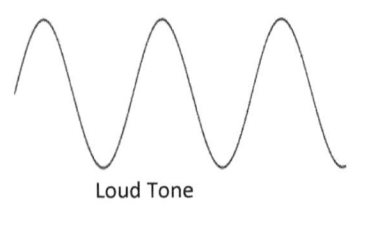
Loud Tone

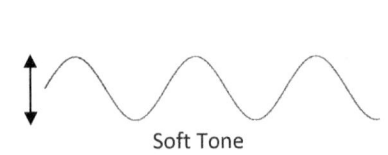
Soft Tone

Quality of wave is affected by waveform

Beats is a difference in frequency of two sound produced at the same time

$$\text{Beats} = |Frequency_1 - Frequency_2|$$

Echo is a reflection of sound wave

Example 10.1

A sound produced by a honk of a truck has a frequency of 3 kHz and a wavelength of 11 cm. a) Find the speed of sound produced in m/s
   b) If the cliff was 500 meter away from the car, how long will it take the the car t hear an echo ?

solution: a) $f = 3000$ Hz, $\lambda = 11 \times 10^{-2}$ m

   $v = f \times \lambda = 3000 \times 11 \times 10^{-2} =$ __330 m/s__

   b)

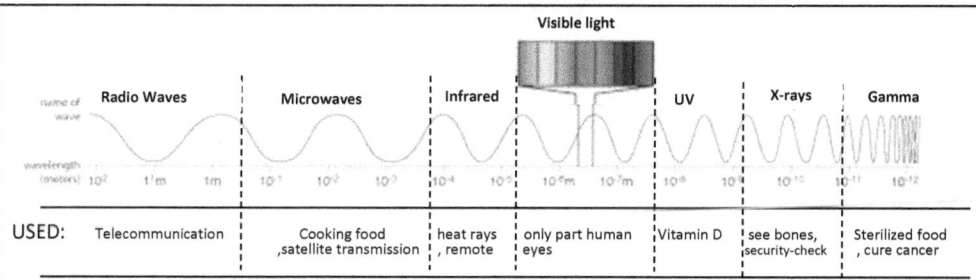

We use speed = distance / time

time = distance / speed

time = (going distance + coming distance) ÷ speed

time = (500 + 500) ÷ 330 = __3.03 seconds__

Electromagnetic Wave

name of wave	Radio Waves	Microwaves	Infrared	Visible light	UV	X-rays	Gamma
wavelength (meters)	10^2 1 m 1 m	10^{-1} 10^{-2} 10^{-3}	10^{-4} 10^{-5}	10^{-6} m 10^{-7} m	10^{-8} 10^{-9}	10^{-10} 10^{-11}	10^{-12}
USED:	Telecommunication	Cooking food, satellite transmission	heat rays, remote	only part human eyes	Vitamin D	see bones, security-check	Sterilized food, cure cancer

Properties of Electromagnetic Waves:

1. Travel at the same speed (= 3×10^8 m/s)
2. All are transverse wave and carry energy
3. All can travel in vacuum (need no medium)
4. All can be reflected, refracted and diffracted
5. High frequency = High Energy

Mechanical Waves

Wave that needs a medium to carry energy from one point to another point. Examples are water wave, sound wave and seismic wave.

Reflection

Refelction of wave (bouncing) occurs when a wave hits a surface of a shinny or a solid object.

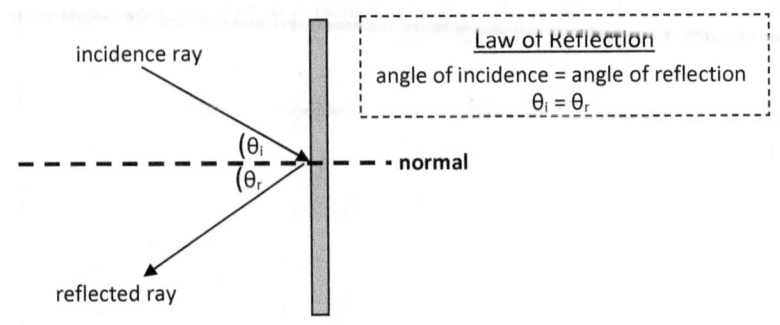

Law of Reflection
angle of incidence = angle of reflection
$\theta_i = \theta_r$

Refraction

Refratcion of wave (bending) occurs when wave speed changes.

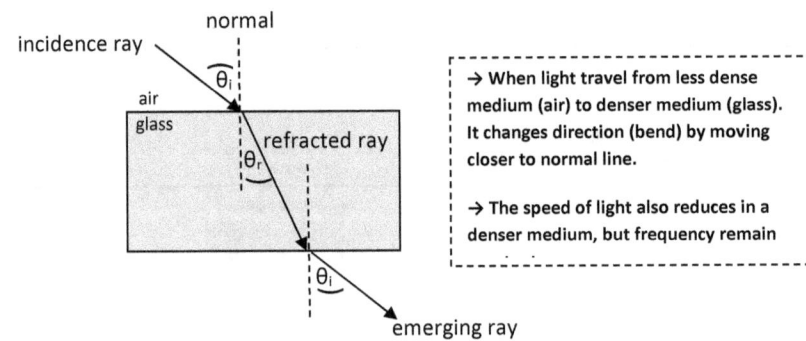

→ When light travel from less dense medium (air) to denser medium (glass). It changes direction (bend) by moving closer to normal line.

→ The speed of light also reduces in a denser medium, but frequency remain

Snell's law of refraction

medium 1   medium 2

$n_1 \sin(\theta_1) = n_2 \sin(\theta_2)$

$n_1 v_1 = n_2 v_2$

Law of refraction

$n = \dfrac{\sin(\theta_i)}{\sin(\theta_r)} = \dfrac{v_i}{v_r}$

n → refractive index

v → speed of wave or speed of light

$n_{air} = 1$

θ_i → angle of incidence

θ_r → angle of refraction

$v_{light\ in\ air} = 3 \times 10^8$ m/s

Total internal reflection

Case1: only refaction occurs   Case2: refraction at 90°   Case3: **total internal reflection**

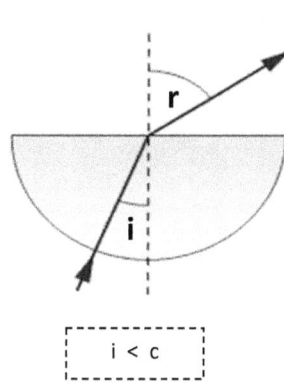
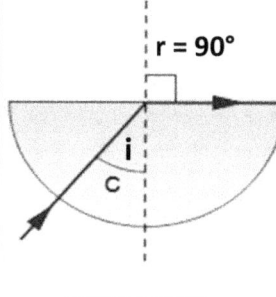
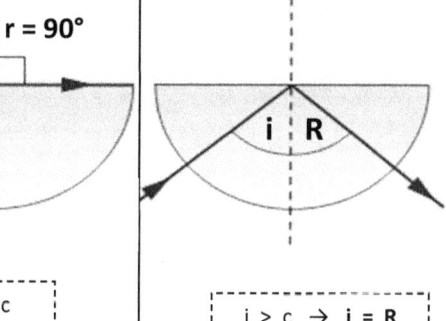

i < c

angle of incidence is less than critical angle

i = c

angle of incidence is equal to critical angle

i > c → i = R

if angle of incidence is greater than critical angle then total internal reflection will occur

$$n = \frac{1}{\sin(c)}$$

n → refractive index

c → critical angle

Use of total internal reflection:

FIBER OPTIC is used in data transmission by manipulating light as a signal.

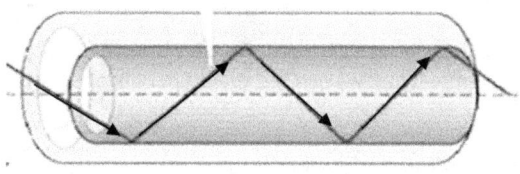

Diffraction

Diffraction of wave (spreading) occurs when wave move thru a narrow gap.

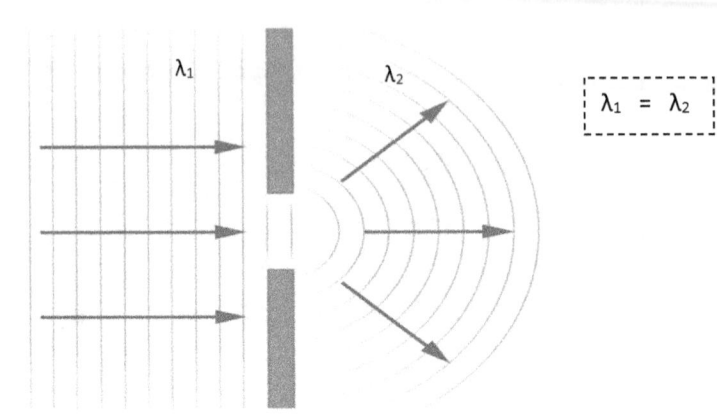

$\lambda_1 = \lambda_2$

Dispersion

Dispersion of wave (visible light disperse to 7 colors) occurs when white light enters a dense medium and give off 7 colors

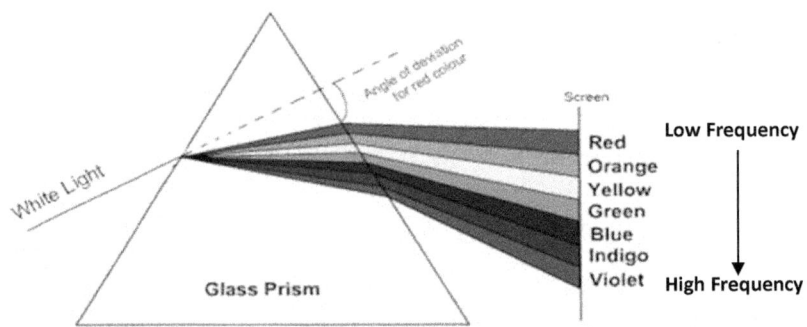

→ When white light travel in glass prism its speed reduces, we can see spectrum of light coming out

→ Violet leaves the prism first, because of its frequency is highest (highest energy)

Mirror

Plane mirror

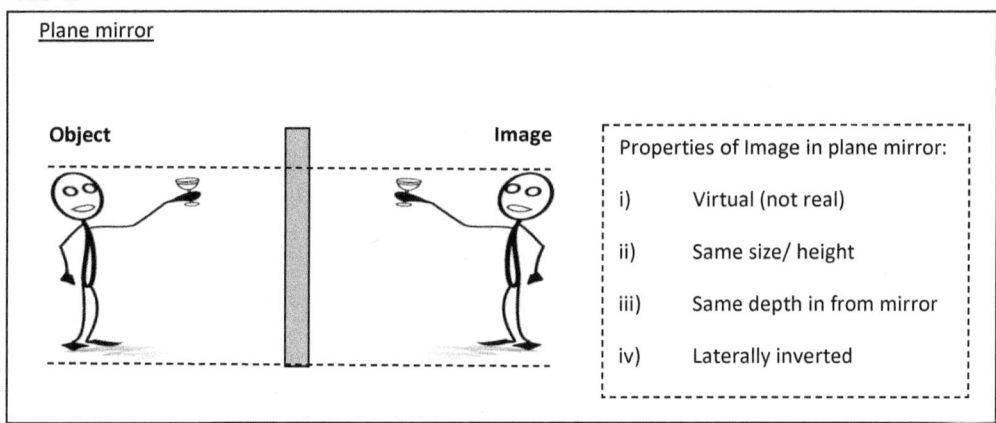

Properties of Image in plane mirror:

i) Virtual (not real)

ii) Same size/ height

iii) Same depth in from mirror

iv) Laterally inverted

Concave mirror

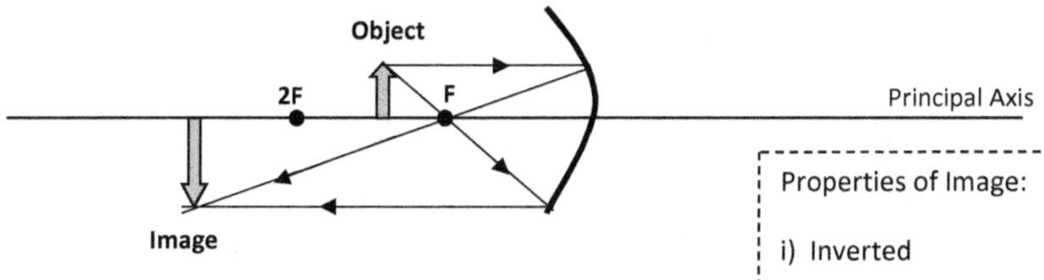

Properties of Image:

i) Inverted

ii) Magnified and Real

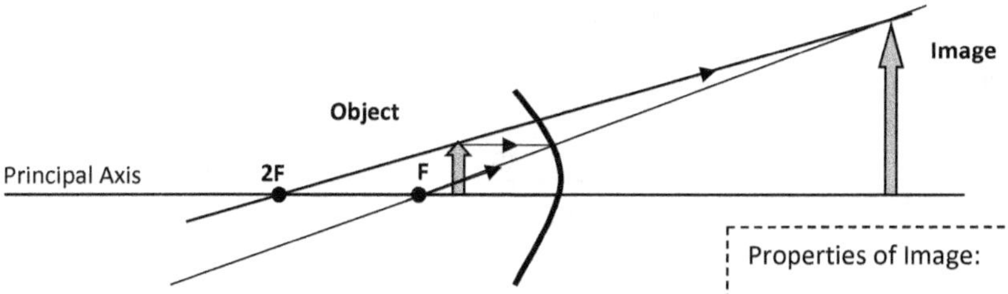

Properties of Image:

i) Laterally inverted

ii) Magnified and Virtual

Convex mirror

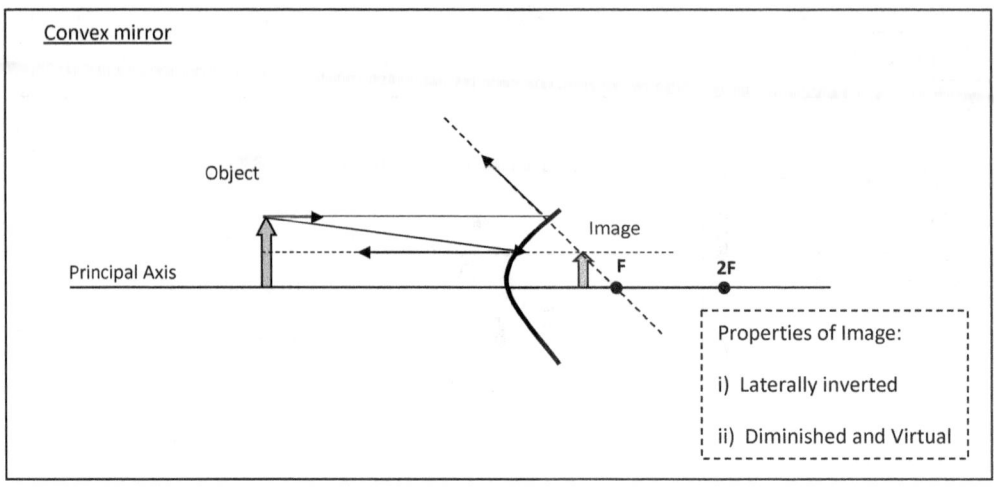

Properties of Image:
i) Laterally inverted
ii) Diminished and Virtual

Formula for curvered mirror :

$$\frac{1}{focus} = \frac{1}{dist\ of\ object} + \frac{1}{dist\ of\ image}$$

$$magnification\ (M) = \frac{Height\ of\ image}{Height\ of\ object} = \frac{dist\ of\ image}{dist\ of\ object}$$

If M < 1 then image is diminished (smaller)

If M > 1 then image is magnified (bigger)

If M = 1 then image is same size as object

Radius of curvature = 2 x focus

SAT Physics (Physics made simple)

Lens

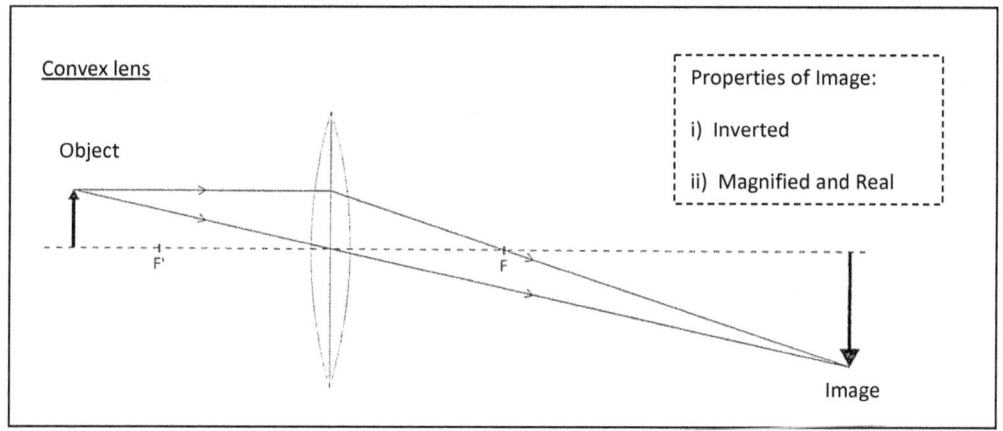

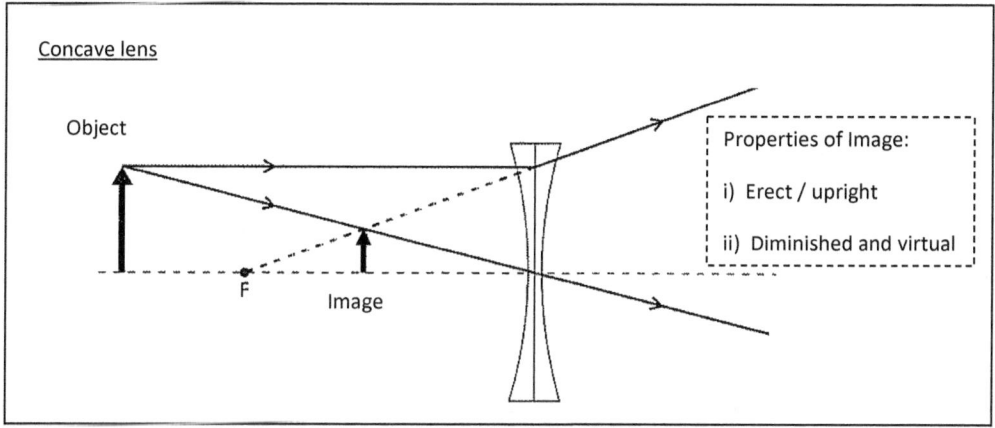

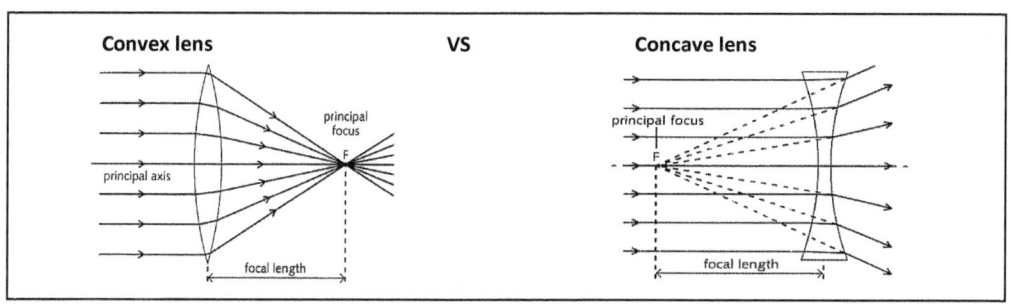

Formula for lens

$$\frac{1}{focus} = \frac{1}{dist\ of\ object} + \frac{1}{dist\ of\ image}$$

$$magnification\ (M) = \frac{Height\ of\ image}{Height\ of\ object} = \frac{dist\ of\ image}{dist\ of\ object}$$

If M < 1 then image is diminished (smaller)

If M > 1 then image is magnified (bigger)

If M = 1 then image is same size as object

Practice Chapter 10

1. A dog is howling at night at the frequency of 30 kHz, the owner was not able to hear. What is a best conclusion for this ?

   a) a wave produce has a high amplitude

   b) a wave produce is an electromagnetic wave

   c) a wave is a longitudinal wave

   d) a wave produced is beyond human audible range

   e) a wave produced is not loud

2. During a storm a buoy is seen moving from crest to trough in 20 millisecond, what is the frequency of the wave produced ?

   a) 25   Hz

   b) 50   Hz

   c) 100   Hz

   d) 200   Hz

   e) 400   Hz

3. Jacob stands 2.5 meter from the mirror, the distance between his image and him is

   a) 1.25   m

   b) 2.0   m

   c) 2.5   m

   d) 4.0   m

   e) 5.0   m

4. A wave has a pitch of 400 Hz and a wavelength of 3 cm, what is the speed of this wave ?

   a) 12   cm /s
   b) 120   cm/s
   c) 12   m/s
   d) 120   m/s
   e) 1200   m/s

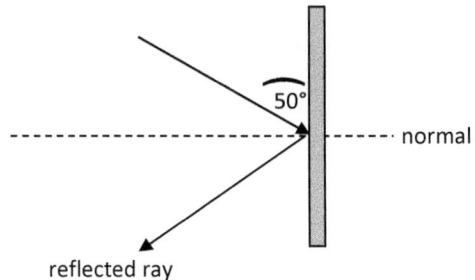

5. According to the diagram above, what is the angle of reflection ?

   a) 30°
   b) 40°
   c) 50°
   d) 60°
   e) 100°

6. Which phenomena cannot occur with blue light of single wavelength ?

   a) dispersion
   b) reflection
   c) diffraction
   d) refraction
   e) polarization

Question 7 - 9

A boy 150 cm tall stand in front of concave mirror, with a focal length of 50 cm. The boy stands 30 cm from the mirror.

7. The image produce by the mirror will be how far from the boy ?

   a) 15   cm
   b) 30   cm
   c) 45   cm
   d) 75   cm
   e) 105   cm

8. What are the property of image produced ?

   a)   Real and magnified
   b)   Virtual and magnified
   c)   Real and diminished
   d)   Virtual and diminished
   e)   Virtual and inverted

9. How tall is the image produced ?

   a) 15   cm
   b) 30   cm
   c) 45   cm
   d) 75   cm
   e) 105   cm

10. Which of the following is not true about electromagnetic wave ?

    a)   all travel at speed of 3×10^8 m/s
    b)   can be refracted
    c)   small wavelength waves are low in energy
    d)   behave like a transverse wave
    e)   low frequency waves are contain less energy

11. When light leaves the water into the air, which of the following is true ?

   a)   velocity increases, wavelength increases and frequency stay same

   b)   velocity increases, wavelength increases and frequency decreases

   c)   velocity increases, wavelength decreases and frequency increases

   d)   velocity decreases, wavelength decreases and frequency stay same

   e)   velocity decreases, wavelength decreases and frequency decreases

Question 12 -13

Visible light enters the glass that has a refractive index of 1.47

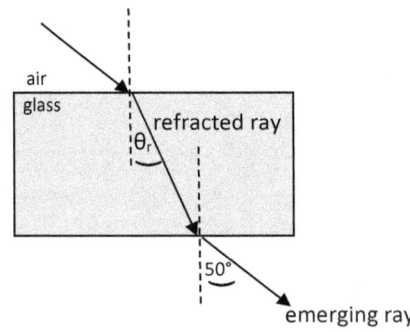

12. What is the angle of refraction ?

   a)   15.6°

   b)   20.3°

   c)   24.6°

   d)   31.4°

   e)   42.5°

13. What is the speed of light in this glass ?

   a) 4.41×10^8 m/s

   b) 4.02×10^8 m/s

   c) 3×10^8 m/s

   d) 2.30×10^8 m/s

   e) 2.04×10^8 m/s

14. An object 5 mm tall viewed under magnifying glass has an image of 10 cm. What is the magnification size of image produced by a convex lens?

   a) 2

   b) 5

   c) 10

   d) 15

   e) 20

15. A bat navigate using sonar pulse with speed of 343 m/s, if the bat cave was 200 meters away from the bat how long will it take pulse to travel complete cycle ?

   a) 1.25   s

   b) 1.20   s

   c) 1.17   s

   d) 0.58   s

   e) 0.42   s

Answer: Chapter 10

Question	Answer	Explanation
1.	D	Human audible range are between 20 to 20,000 Hz So the howl is 30,000 Hz, which is beyond human audible range
2.	A	The period of the wave here is 40 ms Frequency = 1 ÷ period Frequency = 1 ÷ (40 x 10^{-3}) = 25 Hz
3.	E	The question for ask this length, which is 5 m
4.	C	v = f x λ v = 400 x 0.03 v = 12 m/s
5.	B	Angle of incidence = 40° Angle of reflection = angle of incidence Angle of reflection = 40°
6.	A	Dispersion can occur with only white light thru prism, glass or vapor
7.	E	$\frac{1}{50} = \frac{1}{30} + \frac{1}{D_i}$ D_i = -75 cm 75 cm inside the mirror How far from the boy ? 30 + 75 = <u>105 cm</u>

SAT Physics (Physics made simple)

Question	Answer	Explanation
8.	B	Image produce here gives a negative value (-75) which means it is bigger and also inside the mirror, therefore it virtual and magnified
9.	D	$\dfrac{1}{50} = \dfrac{1}{150} + \dfrac{1}{H_i}$ $H_i = 75$ cm Image produce will be 75 cm tall
10.	C	All of the following are true except 'C' small wavelength waves means high frequency (= high in energy)
11.	A	When light travel from denser to less dense it will move faster So Velocity increases, frequency always remains constant then wavelength would increase

12.	D	According to Snell's law
		   Air Glass
		$n_1 \sin(\theta_1) = n_2 \sin(\theta_2)$
		$1 \times \sin(50) = 1.47 \sin(r)$
		       r = 31.4°

13.	E	Air Glass $n_1 V_1 = n_2 V_2$ $1 \times (3.10^8) = 1.47 \times V$ V $= 2.04 \times 10^8$ m/s
14.	E	magnification = $\dfrac{\text{Height of image}}{\text{Height of object}} = \dfrac{10 \times 10^{-2}}{5 \times 10^{-3}} = 20$
15.	C	Speed = distance ÷ time → time = distance ÷ speed Time = (200 + 200) ÷ 343 = 1.17 seconds

Wave II   11

Check-list:

- ✓ *Simple harmonic motion*
- ✓ Standing waves
- ✓ Illumination & Polarization
- ✓ Doppler effect
- ✓ Interference of Wave & Double slits

Simple Harmonic Motion(SHM)

Simple harmonic motion is a motion that repeats itself until all its energy is used up, transformed to heat or sound. Such wave can be demonstrated by a vibrational movement of spring or an oscillations of a pendulum.

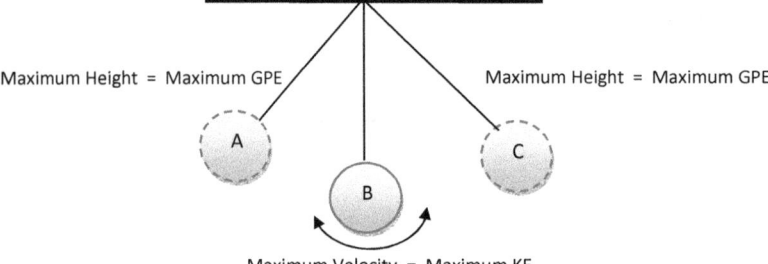

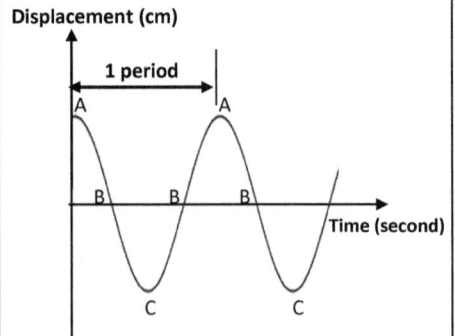
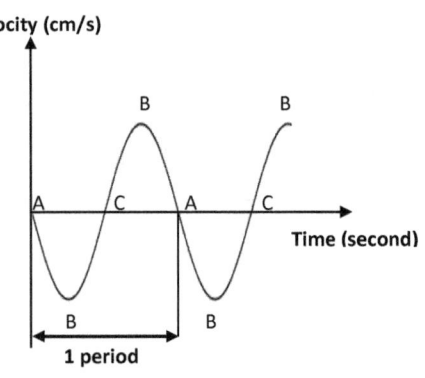

Period of SHM

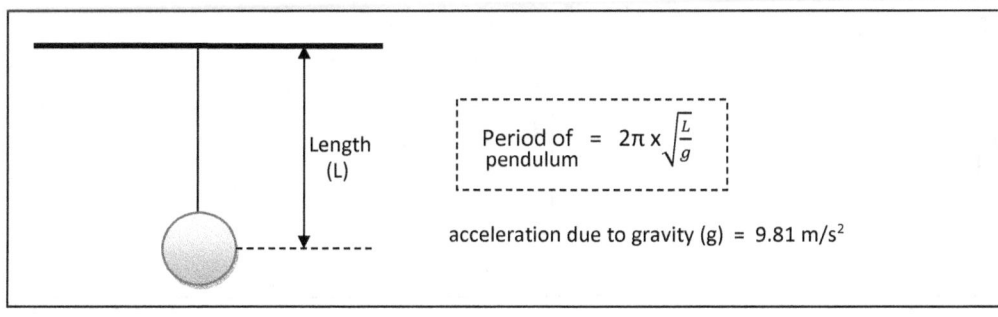

Period of pendulum = $2\pi \times \sqrt{\dfrac{L}{g}}$

acceleration due to gravity (g) = 9.81 m/s²

Period of spring = $2\pi \times \sqrt{\dfrac{m}{k}}$

'k' is spring-constant

Example 11.1

Mike wants a pendulum to oscillate with a period of T second so he uses a pendulum of length L meter, somehow Mike discovered that his pendulum has a period of T/2 seconds. By what factor must he change the length of his pendulum ?

Solution: Period = $2\pi \times \sqrt{\dfrac{L}{g}}$ → $T = K \times \sqrt{L}$ (Let $K = \dfrac{2\pi}{\sqrt{g}}$)

$$\dfrac{T1}{\sqrt{L1}} = \dfrac{T2}{\sqrt{L2}}$$

$$\dfrac{T/2}{\sqrt{L}} = \dfrac{T}{\sqrt{??}}$$

$$\sqrt{??} = 2 \times \sqrt{L}$$

$$L_{new} = 4 \times L$$

New Length is 4 times of old Length (L)

Standing waves

When a wave is created within a boundaries where it can bounce back and forth, such as string tied between two poles. Standing wave are also called stationary wave.

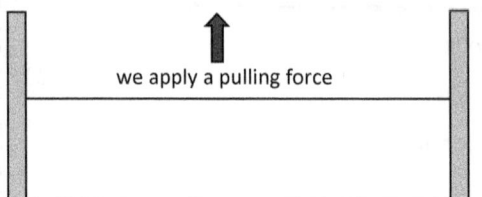

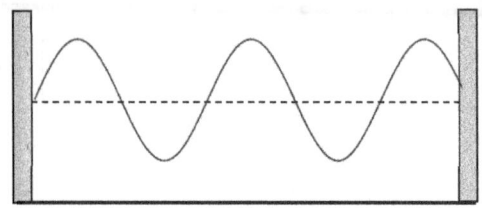

a standing wave with a fundamental frequency

Fundamental frequency lowest frequency produced by free oscillating object

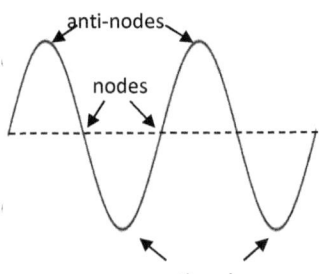

Anti-nodes is a point along a standing wave that has maximum displacement

Node is a point along a standing wave where displacement is zero

$$\text{Wave speed} = \sqrt{\frac{\text{Tension on the string}}{\text{Linear mass density}}}$$

$$\text{Linear mass density} = \frac{\text{total mass}}{\text{total length}}$$

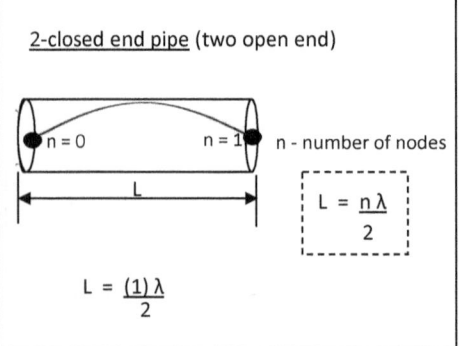

2-closed end pipe (two open end)

n - number of nodes

$L = \frac{n\lambda}{2}$

$L = \frac{(1)\lambda}{2}$

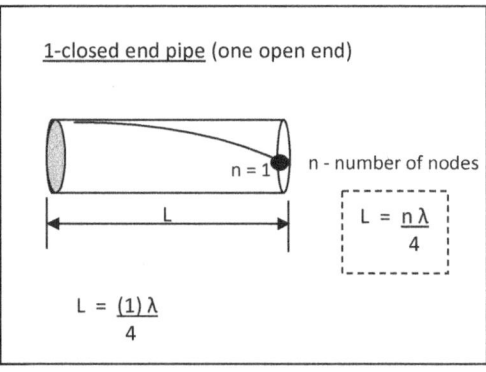

1-closed end pipe (one open end)

n - number of nodes

$L = \frac{n\lambda}{4}$

$L = \frac{(1)\lambda}{4}$

SAT Physics (Physics made simple)

Illumination

Intensity of illumination produced by a point source varies inversely as square of a the distance from the source

$$\text{Intensity} = \frac{1}{distance^2}$$

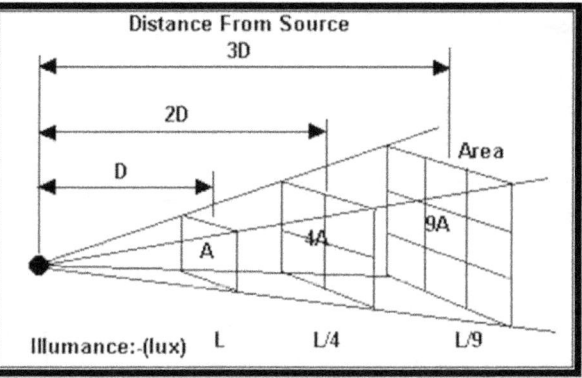

Credit: www.cctv-information.co.uk

Polarization

Electromagnetic waves behave like a transverse wave, since they are oscillating in more than one plane. So we can polarize this wave by limiting its direction of motion on to one plane using a polaroid. **Polaroid** are crystal sheets or synthetic material made from plastics that can filter (restrict the direction of) oscillation in transverse wave.

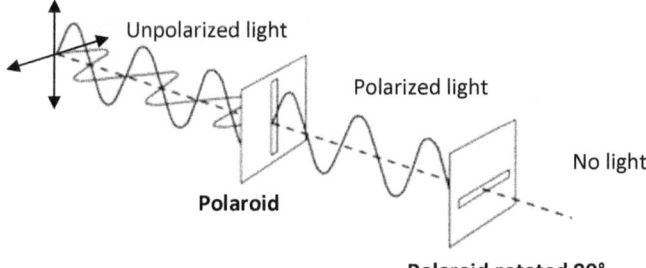

142 | Page

SAT Physics (Physics made simple)

Maxwell's theory said light is a transverse wave where electric field(E) and magnetic filed(B) vibrate at right angles to direction of motion

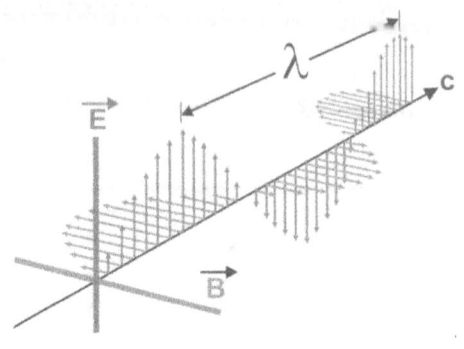

Doppler effect

If a source of sound is moving toward an observer A the frequency of the sound observe by him will be higher than original one. While observer B will observe a frequency that is much lower than the original one.

Moving source

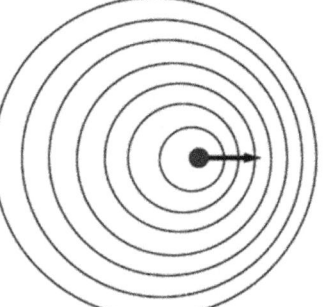

observer B

observer A

See:
Low Frequency
Longer Wavelength

$$f = \frac{f_0 \cdot V}{V + V_s}$$

See:
High Frequency
Shorter Wavelength

$$f = \frac{f_0 \cdot V}{V - V_s}$$

f - observed frequency f_0 - transmitted frequency

V - wave velocity V_s - source velocity

Moving observer

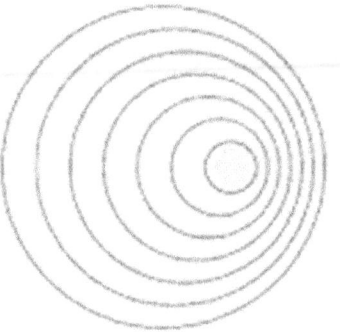

observer B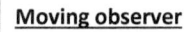

See:
Low Frequency
Longer Wavelength

$$f = f_0 \frac{(V - V_s)}{V}$$

observer A

See:
High Frequency
Shorter Wavelength

$$f = f_0 \frac{(V + V_s)}{V}$$

f - observed frequency

V - wave velocity

f_0 - transmitted frequency

V_s - source velocity

Interference of Wave

Constructive Interference occurs when two waves that are in phase collide with each other to form a bigger wave.

 +

Destructive Interference occurs when two waves that are out of phase collide with each other and canceled some part out or no wave.

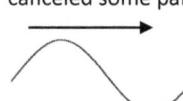 +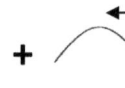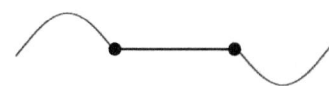

Shock wave is created when a moving object breaks the sound barrier, travel faster than its sound.

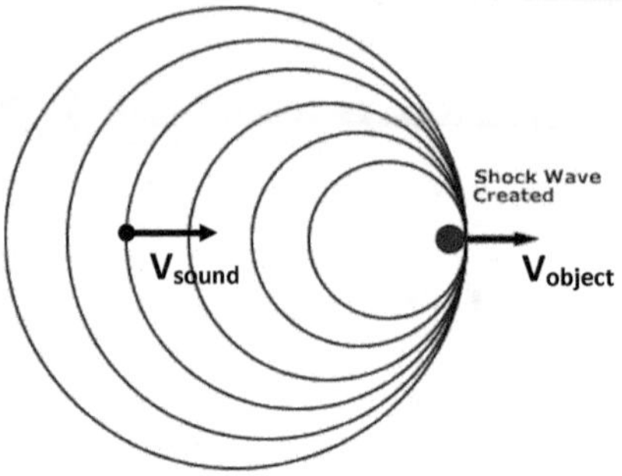

V_{sound} Shock Wave Created V_{object}

Thomas Young double slits

Thomas Young discovered that light waves interfere with each other in such a way that constructive(bright spot) and destructive(dark spot) interference appear to occur and this can be shown on the screen.

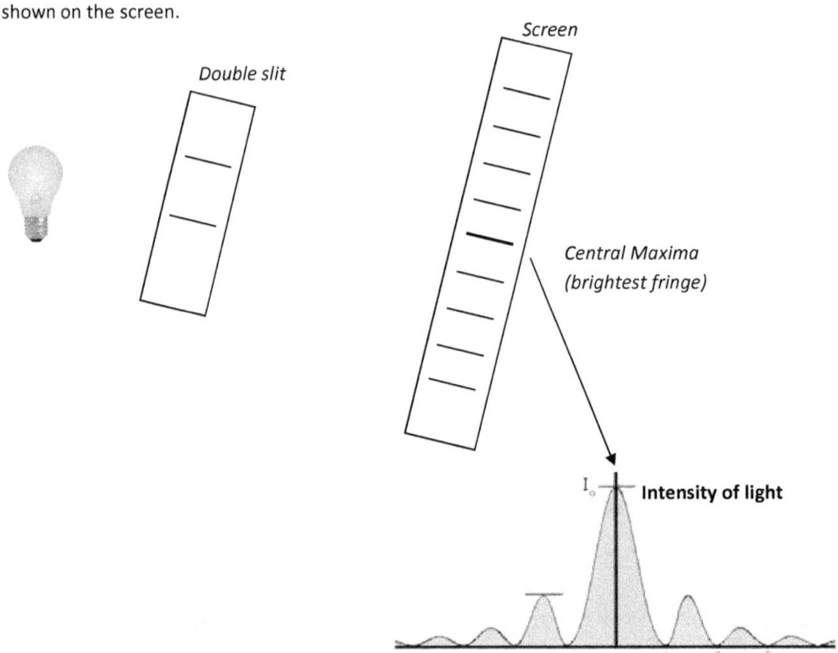

Double slit

Screen

Central Maxima (brightest fringe)

I_o Intensity of light

Top view from double slit

n =3　　　n=2　　　n=1　　　↑ n =0　　　n =1　　　n=2　　　n=3
Central maxima

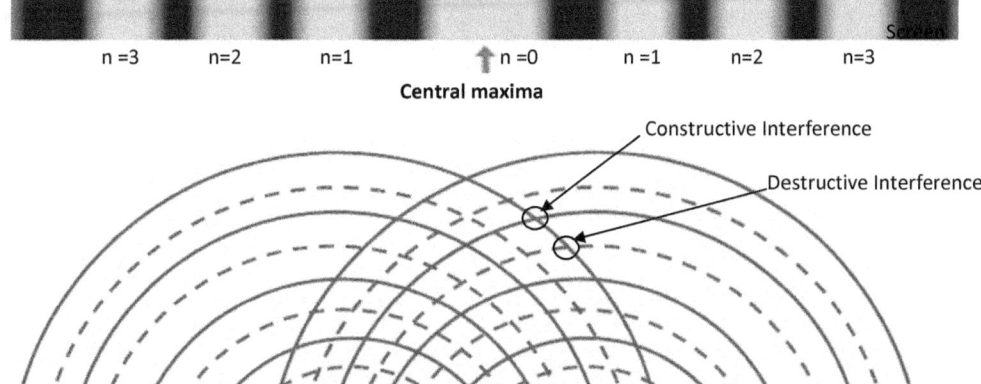

Interference of the waves from slit 1 and slit 2 causes bright and dark fringes on the screen

Side view of double slit

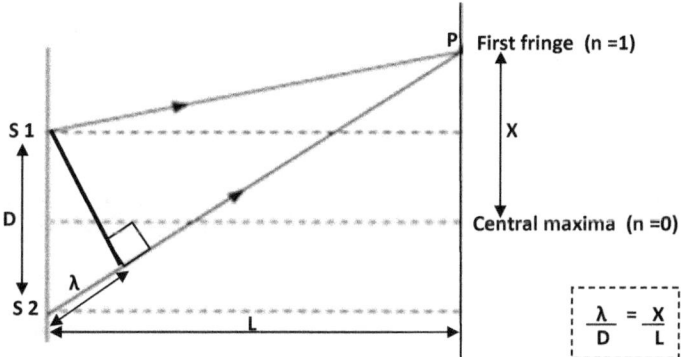

$$\frac{\lambda}{D} = \frac{X}{L}$$

<u>Concept of double slits:</u>

1. Wavelength (λ) is directly proportional to distane of the slits(D)

2. Wavelength (λ) is directly proportional to distane between fringes (X)

3. Wavelength (λ) is indirectly proportional to distance between slits and screen (L)

Practice
Chapter 11

1. When a simple harmonic wave has a maximum displacement, which of following is be true ?

    a) a wave speed is maximum

    b) a wave speed is zero

    c) a wave is a longitudinal wave

    d) a wave acceleration is zero

    e) a wave is not a periodic wave

2. If the length of the pendulum is quadrupled then what happen to the period of wave?

    a) the period is halved

    b) the period is quarter

    c) the period is doubled

    d) the period is quadrupled

    e) the period remain the same

3. The brightness at point A has an intensity of I, if the distance between the lamp and point A is tripled then what is the value of the new intensity ?

    a) 9 x I

    b) 3 x I

    c) 2 x I

    d) I / 3

    e) I / 9

4. Which of the following is true when red light passes through a double slit ?

   a) alternate bands of red and green will be produced

   b) alternate bands of red and yellow will be produced

   c) alternate bands of red and white will be produced

   d) alternate bands of red and black will be produced

   e) alternate bands of black and white will be produced

5. What is the length of a closed end pipe if a tuning fork with frequency of 500 Hz is set to vibrates on it ? Given that the speed of sound in air is 340 m/s

   a)   17   cm

   b)   34   cm

   c)   68   cm

   d)   136   cm

   e)   143   cm

6. Which of the following statement is TRUE ?

   a)   Water wave can be polarized

   b)   Electromagnetic wave are all polarized

   c)   Doppler effect cannot be applied with electromagnetic wave

   d)   Electromagnetic wave are composed of electric and magnetic field in perpendicular plane

   e)   Electromagnetic wave are composed of gravitational and magnetic field in perpendicular plane

 observer B 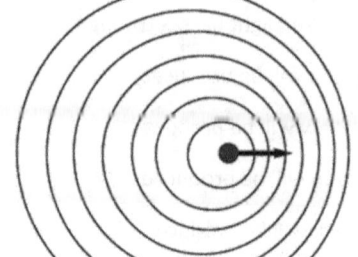  observer A

Question 7 - 9

An ice-cream truck moving with the speed of 0.25V produces a sound with a frequency of 440 Hz and speed of V.

7. What is the frequency that observer A hear ?

    a) 602   Hz

    b) 587   Hz

    c) 556   Hz

    d) 440   Hz

    e) 352   Hz

8. What is the frequency that observer B hear ?

    a) 602   Hz

    b) 587   Hz

    c) 556   Hz

    d) 408   Hz

    e) 352   Hz

9. What is the frequency hear by the truck driver ?

    a) 602   Hz

    b) 587   Hz

    c) 556   Hz

    d) 440   Hz

    e) 352   Hz

10. In Thomas Young double slits experiment, if we double the distance between the slit and the screen then which of the following is NOT TRUE?

   a) wavelength is doubled

   b) distance between fringes is double

   c) distance between the slit should be doubled if wavelength is constant

   d) speed of light is same for both slit

   e) frequency of wave is doubled

11. Given that distance between the slits are 3 cm and the distance from slit to the screen is 30 cm, while distance between the fringes are 2 cm, what is the wavelength of the light ?

   a) 20   mm

   b) 10   mm

   c) 5   mm

   d) 2   mm

   e) 1   mm

Scan me

Answer: Chapter 11

Question	Answer	Explanation
1.	B	Maximum displacement means it has a greatest potential energy so kinetic energy will be lowest. Therefore its speed will be zero.
2.	C	$$\frac{T_1}{\sqrt{L}} = \frac{T_2}{\sqrt{4L}}$$ $$T_2 = T_1 \times 2$$ Period will be doubled if length increased by 4 times.
3.	E	$I_1 d_1^2 = I_2 d_2^2$ $I (1)^2 = I_2 (3)^2$ $I_2 = I/9$
4.	D	Any colored light passes through double slits will produced alternate bands of that color with black color
5.	A	$V = f \times \lambda$ $340 = 500 \times \lambda$ $\lambda = 0.68$ m Since it is a one-closed end we use $L = \lambda/4$ $L = 0.68/4 = 0.17$ m Length is equal 17 cm
6.	D	Electromagnetic wave are composed of electric and magnetic field in perpendicular plane

Question	Answer	Explanation
7.	B	Moving toward observer $\quad f = \dfrac{f_0 \cdot V}{V - V_s} = \dfrac{440 \times V}{V - 0.25V} = \underline{587\ Hz}$
8.	E	Moving away from observer $\quad f = \dfrac{f_0 \cdot V}{V + V_s} = \dfrac{440 \times V}{V + 0.25V} = \underline{352\ Hz}$
9.	D	Doppler effect has no effect on the sound source, so frequency heard by the driver will be same as frequency emitted of 440 Hz
10.	A	According to Young If 'L' → 2L Then X → 2X $\quad\quad\quad$ D → 2D $\quad\quad\quad$ λ → 0.5λ So only wrong choice here is A since wavelength should be halved
11.	D	$\dfrac{\lambda}{D} = \dfrac{X}{L} \rightarrow \dfrac{\lambda}{3} = \dfrac{2}{30} \rightarrow \lambda = 0.2\ cm = \underline{2\ mm}$

Gravitational and Electric field   12

Check-list:

- ✓ Kepler's Law of planetary motion
- ✓ Newton's Law of Gravitational
- ✓ Gravitational Potential Energy
- ✓ Electric Field and Potential

Kepler's law of planetary motion

Kepler's law can be summarized as follow:

1. All planets orbit around the Sun in an elliptical path

2. A planet sweeps out equal amount of area within equal amount of time
   (conservation of angular momentum)

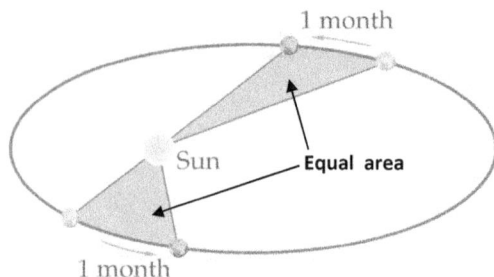

(When a planet is nearer to the Sun it will move faster)

3. The ratio of a planet mean distance from the Sun (R) cube to the orbital period around the Sun(T) square is equal to Kepler's constant (K_s)

$$K_s = \frac{R^3}{T^2}$$
$$K_s = 3.35 \times 10^{18} \ m^3 s^{-2}$$

Newton's law gravitational

In Universe, two objects exert an attractive force on each other known as **gravitational force(F_g)**

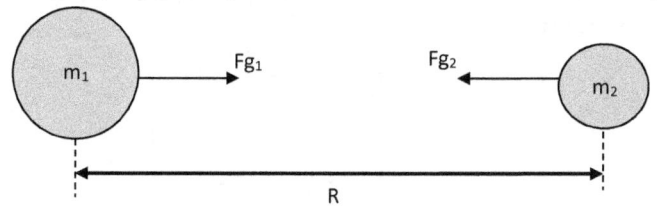

$$F_{g1} = F_{g2} = F_g = \frac{G m_1 m_2}{R^2}$$

where 'G' is gravitational constant
$G = 6.67 \times 10^{-11}$ N m² kg⁻²

m → mass (kg)

R → mean distance (m)

Concepts:

1. Force is directly proportional to the product of the masses
2. Force is inversely proportional to the radius square

We can relate Newton's law of gravitational with Newton's law of motion to obtain the value of acceleration due to gravity, given that we know the mean radius and mass of the planet.

$$F = ma \longleftrightarrow F = \frac{G m_1 m_2}{R^2}$$

$$\cancel{m} a = \frac{G \cancel{m_1} m_2}{R^2}$$

$$a_g = \frac{G m_{planet}}{R^2}$$

$M_{Earth} = 6 \times 10^{24}$ kg

$R_{Earth} = 6.378 \times 10^6$ m

$a_g = 9.81$ m/s²

Example 12.1

A distant planet has half the mass and half the radius of the Earth. What is the acceleration due to gravity of this planet ?

Solution: We can compare the acceleration due to gravity of both planet by

Earth → $a_{gE} = \frac{G M_{Earth}}{R^2_{Earth}}$ → $a_{gE} = 9.81$ m/s²

Distant planet → $a_{gP} = \frac{G(0.5 M_{Earth})}{(0.5 R_{Earth})^2}$ → $a_{gP} = 2 \times \frac{G M_{Earth}}{R^2_{Earth}}$ → $a_{gP} = 2\, a_{gE}$

$a_{gP} = 2 \times 9.81 = \underline{19.62\ m/s^2}$

Gravitational Potential Energy

As we move away from the Earth the potential energy exerted on us will generally decrease. We can calculate amount of energy exerted upon us by using:

$$\text{Work} = \Delta GPE = \Delta U$$

$$F_g \times R = \frac{G\, m_1\, m_2}{R^2} \times R$$

$$\Delta U = \frac{G m_{object} \times m_{Earth}}{R}$$

To calculate the escaping velocity from any planet we can equate the potential energy to its kinetic energy.

$$KE = \Delta GPE$$

$$\frac{m V^2_{esapce}}{2} = \frac{G\, m\, m_e}{R^2_e}$$

$$V_{escape} = \sqrt{\frac{2GM_e}{R_e}}$$

While the orbital velocity can be found by equating centripetal force with gravitation force.

$$F_c = F_g$$

$$\frac{m V^2}{R} = \frac{G\, m_1\, m_2}{R^2}$$

$$V_{orbit} = \sqrt{\frac{GM_e}{R_e}}$$

Example 12.2

A distant planet has half the mass and half the radius of the Earth. What is the escaping velocity from this planet?

Solution: $\quad V_{escape} = \sqrt{\frac{2G\,0.5M_e}{0.5 R_e}} = \sqrt{\frac{2GM_e}{R_e}}$

Plugging in the value we get V_{escape} = <u>11,200 m/s</u>

Electric field

In space, two charges exert a force on each other known as **electric force**(F_e)

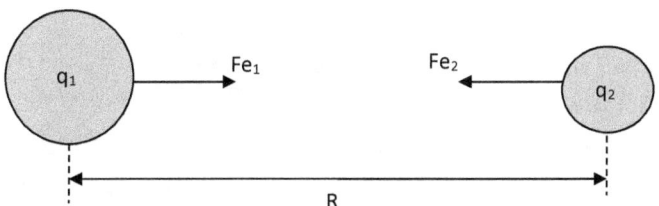

$$F_{e1} = F_{e2} = F_e = \frac{K q_1 q_2}{R^2}$$

where 'k' is coulomb constant
k = 8.99 x 10^9 N $m^2 C^{-2}$

q → charge (Coulomb)

R → mean distance (meter)

Concepts:

1. Force is directly proportional to the product of the charges
2. Force is inversely proportional to the radius square

Charges

Charge define electric properties of proton(positive) or electron(negative), it also states the magnitude. Like charges will repel each other but opposite charges will attract one another.

A **positive charge** would have an electric field pointing out around its surface

A **negative charge** would have an electric field pointing inside its surface

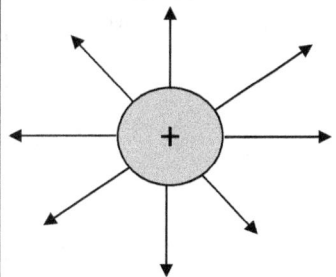
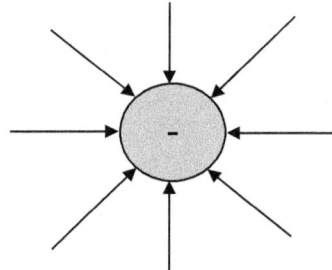

According to law of conservation of charges, total charge in a system will remain constant. Charges flowing a system or circuit at the beginning and ending should be equal.

Electric field between two charges of equal magnitude have the same amount of field line as shown below.

Like charges repelling each other

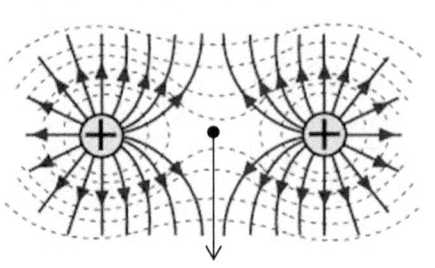

a point where electric field is zero

Opposite charges attracting each other

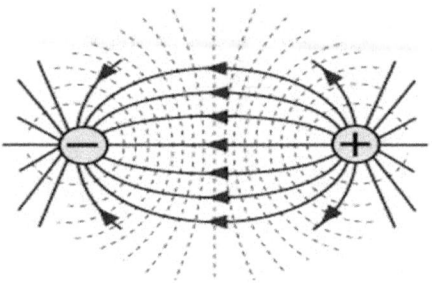

Electric field between two parallel plate

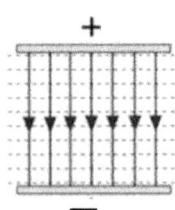

Electric Field can be calculated using formula

Electric Field = Force ÷ charge
$$E \text{ (N/c)} = \frac{F \text{ (N)}}{q \text{ (c)}}$$

We can plug in F_e and rearrange the formula

$$E \text{ (N/c)} = \frac{kq \text{ (c)}}{r^2 \text{ (m)}}$$

Example 12.3

Two positive charges of 5 C and 20 C are placed a distant 12 meter apart. At which point between them would the electric field be equal to zero ?

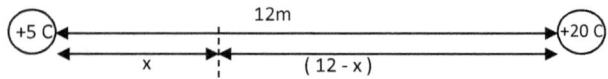

Solution: Let's assume that x meter is a point where electric field is zero

$E_1 = E_2$ → $\frac{k(5)}{x^2} = \frac{k(20)}{(12-x)^2}$ → $5(12-x)^2 = 20x^2$ → $(12-x)^2 = 4x^2$ → $12 - x = 2x$ → $x = \underline{4\,m}$

The point where electric field is zero would be 4 meter away from +5 C

SAT Physics (Physics made simple)

Electric Potential

Electric potential is a measure of electrical pressure at a point, measured in volt (V). We can think of electric potential as a push that causes the electrons to flow in a given path. **Potential difference (ΔV)** is a measured of difference between electrical pressure between two points.

Electric-field is uniformly distributed

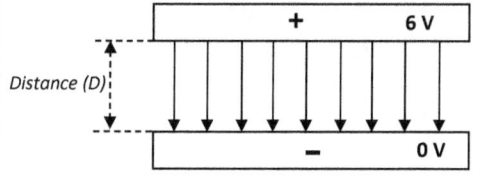

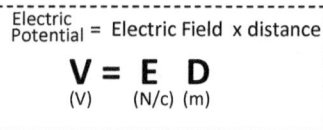

Electric Potential = Electric Field × distance

$$V = E \cdot D$$
$$(V) \quad (N/c) \quad (m)$$

To find potential difference we can use: $\Delta V = E \times \Delta D$

If we insert a charge between the plate of uniform electric field we can calculate its **electric potential energy (U)**

Electric Potential Energy = Charge × electric potential

$$U = q \cdot V$$
$$(J) \quad (c) \quad (V)$$

We can transform the equation above to help us obtain **work done (W)** in moving a charge from one point to another point of different potential.

Work done = Charge × Change in p.d.

$$W = \Delta U = -q \, \Delta V$$
$$(J) \quad (J) \quad (c) \quad (V)$$

Example 12.4

A charge of + 3 C is moved from point X to Y. How much work was done on it?

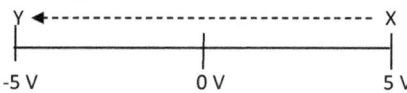

Solution: we can use the $W = -q \Delta V = -q (V_y - V_x) = -(3)(-5-5) = -3 \times -10 = 30 \, J$

The work done in moving a charge from X to Y is equal to 30 Joules

Gravitational VS Electric field

Variables	Gravitational	Electric
Occurs between	Masses (m_1 & m_2)	Charges (q_1 & q_2)
Formula	$F_g = \dfrac{G\, m_1 m_2}{R^2}$	$F_e = \dfrac{K\, q_1 q_2}{R^2}$
Derived into	$F_g = a_c \times m$	$F_e = E \times q$
Potential Energy	$U = -\dfrac{G\, m_1 m_2}{R}$	$U = q\, V$
Type of force	Attractive only	Attractive and Repulsive

Gold leaves electroscope

Gold leaf electroscope is used to identify the charges on an object by looking at the movement of the leaf.

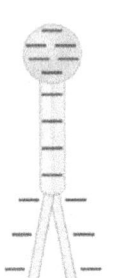

The leaves are negatively charged at start.

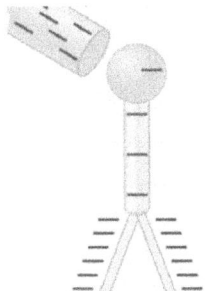

If we bring a negatively charged rod close to the leaves it will **diverge** out from one another, showing that the leaves and rod have the **same charged**.

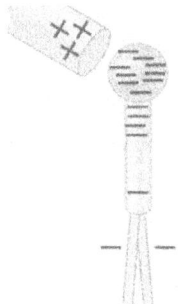

If we bring a positively charged rod close to the leaves it will **converge** in to one another, showing that the leaves and rod have the **different charged**.

Practice
Chapter 12

1. What is the force experienced by two space-crafts of mass 3,000 kg and 5,000 kg that are 0.2km apart ?

    a) 2.5×10^{-8} N

    b) 3.5×10^{-8} N

    c) 4.5×10^{-6} N

    d) 3.5×10^{-6} N

    e) 2.5×10^{-2} N

2. What is the escaping velocity from the moon with a mass of 7.35×10^{22} kilogram and a radius of 1.74×10^{6} meter?

    a) 3373   m/s

    b) 3003   m/s

    c) 2737   m/s

    d) 2374   m/s

    e) 1683   m/s

3. Which of the following is NOT TRUE about Kepler's law of planetary motion ?

    a) All planet orbits the sun in elliptical path

    b) When a planet is far from the Sun it moves slow than when it is close

    c) Ratio of planet's mean radius from the sun square to its period cube is a constant

    d) For a planet of the same mass but different mean distance from the Sun orbital speed are is different

    e) Angular momentum is always conserved

4. What is the orbital speed at the height 3 times the radius of the Earth?

   a) 7921   m/s

   b) 4573   m/s

   c) 3960   m/s

   d) 2743   m/s

   e) 1683   m/s

5. A planet called RB11 has the same gravitational field as the Earth, given that the mass of this planet is 2.3 x 10^{22} kg. What is the approximate radius of this planet?

   a) 4 x 10^5 m

   b) 3 x 10^5 m

   c) 2 x 10^5 m

   d) 1 x 10^5 m

   e) 5 x 10^4 m

6. The ratio potential energy of a satellite orbiting at 'r' compared to a satellite orbiting at '3r' will be

   a) $1 : \sqrt{3}$

   b)   1 : 3

   c)   1 : 9

   d)   3 : 1

   e)   9 : 1

Question 7 - 9

Two massless negatively charges of 3 C and 27 C are placed a distant 13 meter apart.

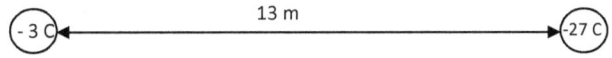

7. What is the electric force between them ?

   a) 5   GN

   b) 4.3   GN

   c) 3   MN

   d) 1.2   kN

   e) 0   N

8. What is the gravitational force between them ?

   a) 5   μN

   b) 4.3   nN

   c) 3.2   pN

   d) 1.2   kN

   e) 0   N

9. At what point approximately from -3 C is the electric field experienced equal zero ?

   a) 0.8   m

   b) 1.3   m

   c) 3.3   m

   d) 4.5   m

   e) 11.7 m

Scan me

10. If the distance is doubled and the charges are doubled by what factor would electric force increased ?

    a) factor of 2

    b) factor of 4

    c) factor of 1/2

    d) factor of 1/4

    e) stay the same

SAT Physics (Physics made simple)

Answer: Chapter 12

Question	Answer	Explanation
1.	B	$F_g = \dfrac{G m_1 m_2}{R^2} = \dfrac{6.67 \times 10^{-11} \times 3000 \times 5000}{200^2}$ $F_g = \underline{2.5 \times 10^{-8} \text{ N}}$
2.	D	$V_{escape} = \sqrt{\dfrac{2GM}{R}} = \sqrt{\dfrac{2 \times 6.67 \times 10^{-11} \times 7.35 \times 10^{22}}{1.74 \times 10^6}} = \underline{2374 \text{ m/s}}$
3.	C	" Ratio of planet's mean radius from the sun **square** to its period **cube** is a constant " is NOT TRUE " Ratio of planet's mean radius from the sun **cube** to its period **square** is a constant " is TRUE
4.	C	$V_{orbit} = \sqrt{\dfrac{GM_e}{R_e}} = \dfrac{(6.67 \times 10^{-11} \times 6 \times 10^{24})^{1/2}}{(4 \times 6.378 \times 10^6)^{1/2}} = \underline{3960 \text{ m/s}}$
5.	A	$a_g = \dfrac{G M_e}{R^2} \rightarrow 9.81 = \dfrac{6.67 \times 10^{-11} \times 2.3 \times 10^{22}}{R^2}$ $R = 395451 \approx \underline{4 \times 10^5 \text{ m}}$
6.	D	$U = -\dfrac{G m_1 m_2}{R} \rightarrow -\dfrac{G m_1 m_2}{r} : -\dfrac{G m_1 m_2}{3r}$ $\phantom{U = -\dfrac{G m_1 m_2}{R} \rightarrow } 3r \ : \ r$ $\phantom{U = -\dfrac{G m_1 m_2}{R} \rightarrow \ } 3 \ : \ 1$
7.	B	$F_e = \dfrac{K q_1 q_2}{R^2} = \dfrac{8.99 \times 10^9 \times -3 \times -27}{13^2}$ $F_e = 4.3 \times 10^9 \text{ N} = \underline{4.3 \text{ GN}}$

Question	Answer	Explanation
8.	E	Since the charges are massless, the gravitational field would occur if there are masses between two bodies. In this case, gravitational field between two bodies is zero due to the zero mass.
9.	B	$$E_1 = E_2$$ $$\frac{k(-3)}{x^2} = \frac{k(-27)}{(13-x)^2}$$ $$3(13-x)^2 = 27 x^2$$ $$(13-x)^2 = 9 x^2$$ $$13 - x = 3x$$ $$x = \underline{3.3 \text{ m}}$$
10.	E	$$F_e = \frac{K q_1 q_2}{R^2} = \frac{8.99 \times 10^9 \times -6 \times -54}{26^2}$$ $$F_e = 4.3 \times 10^9 \text{ N} = \underline{4.3 \text{ GN}} \leftarrow \text{this value is } \textbf{same as before}$$

Electricity   13

Check-list:

- ✓ Ohm's Law
- ✓ Circuit Symbol
- ✓ AC and DC
- ✓ Series VS Parallel
- ✓ Other components

Ohm's Law

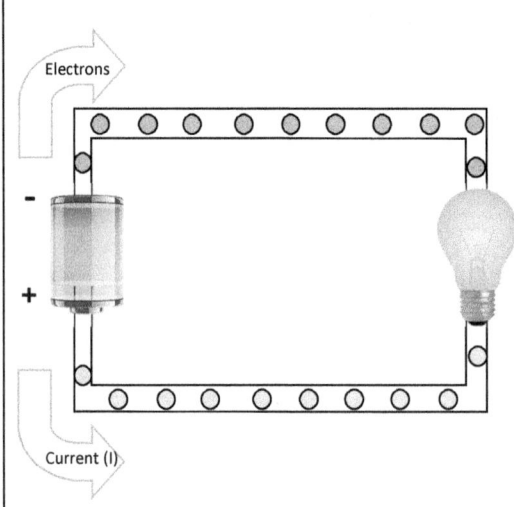

Ohm's Law state that if we varies the current flow the potential difference will be directly proportional

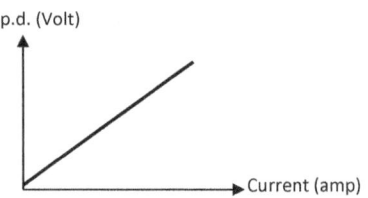

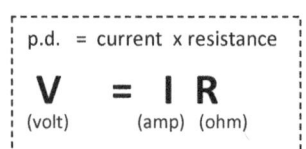

p.d. = current x resistance

$$V = I\ R$$
(volt) (amp) (ohm)

Electrons flow from "-" to "+"

Current flow from "+" to "-"

p.d. (potential difference) is a difference in electrical pressure between positive and negative terminal of the battery that causes electrons to flow out, its unit is Volt.

Current is a measure of rate of flow of charges in a circuit, its unit is Ampere.

Resistor is used to attract electrons out of the cell, its unit is Ohm (Ω).

From previous chapter we have learned that charges can flow, we can measure the rate of flow by using formula below.

$$\text{current} = \text{charges flow} \div \text{time}$$
$$I\ (\text{amp}) = \frac{\Delta Q\ (\text{coulomb})}{\Delta t\ (\text{second})}$$

Example 13.1

In a circuit 30 coulomb of charge flow in 5 seconds through a 10 Ω resistor.

a) Find current in this circuit   b) Find the potential difference across the resistor

Solution: a) We use $I = \frac{Q}{t} = \frac{30}{5} =$ **6 amperes**

b) We apply Ohm's Law

$$V = IR$$

$$V = 6 \times 10 = \underline{60\ Volt}$$

Internal resistance is a small resistor placed inside a battery, to attract electrons back to the battery.

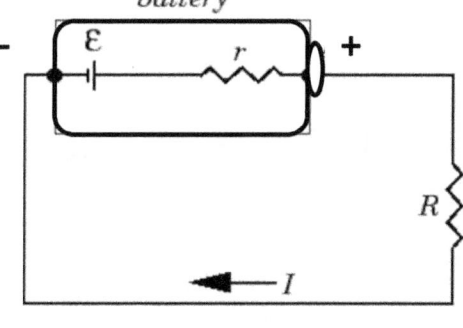

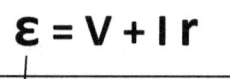

$$\varepsilon = V + Ir$$

e.m.f.(electro motive force) is the maximum electrical pressure or p.d. that a cell has

We can derived and get the formula to find current here by:

$$\varepsilon = IR + Ir$$
$$\varepsilon = I(R + r)$$
$$I = \frac{\varepsilon}{(R + r)}$$

SAT Physics (Physics made simple)

Circuit Symbol

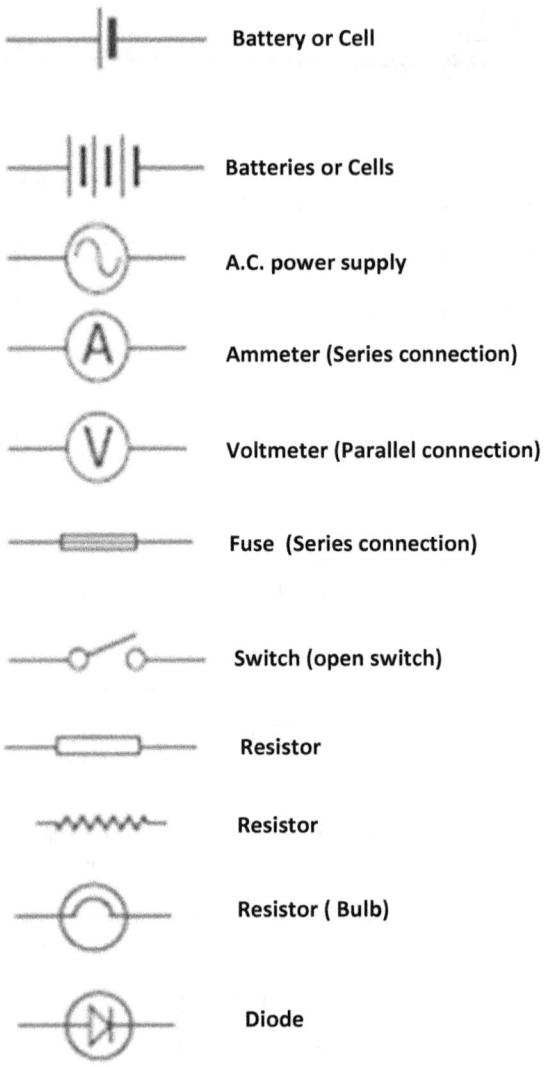
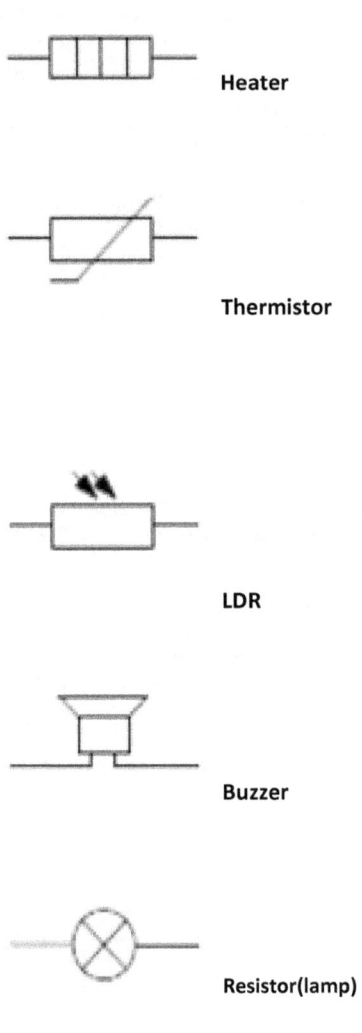

Example 13.2

A battery with e.m.f. of 20 volt is connected to the bulb with a resistance of 8 Ω, given that the internal resistance (r) is 2 Ω. Find a) Reading on ammeter b) p.d. across the bulb

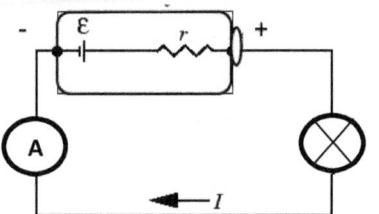

solution: a) we use $I = \varepsilon \div (R + r)$

$I = 20 \div (8 + 2)$

$I = \underline{2\,A}$

Reading on ammeters will be 2 amperes

b) to find p.d. across the bulb we use Ohm's Law :

$V = IR$

$V = 2 \times 8 = \underline{16\,v}$

Factors affecting the resistance of a resistor

1. Resistance(R) is directly proportional to Length (L)
2. Resistance(R) is inversely proportional to Surface Area (A)
3. Resistance(R) increases as temperature increase (ρ - resistivity)
4. Resistance(R) depend on the material itself (ρ - resistivity)

$$R = \frac{\rho L}{A}$$

for example :

5 meter = 20 Ω

2.5 meter = 10 Ω

A.C. and D.C.

Alternating Current (a.c.) is a current that reverse direction alternately from positive to negative and negative to positive.

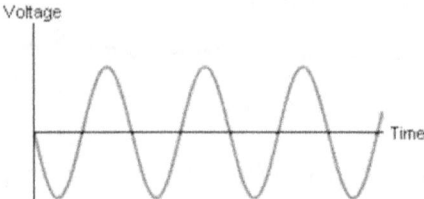

concepts:
1. Home plug socket are all a.c.
2. Obtained directly from generator
3. Used in transmission line to transmit electricity (low heat loss on wire)

Direct Current (d.c.) is a current that flows in one direction only (either positive or negative)

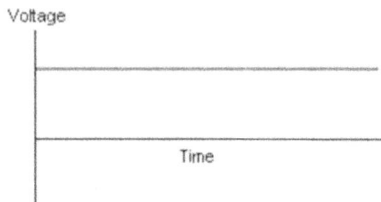

concepts:
1. Batteries and cells are considered to be d.c.
2. Used in every electronic devices such as mobile phone, radio and etc.

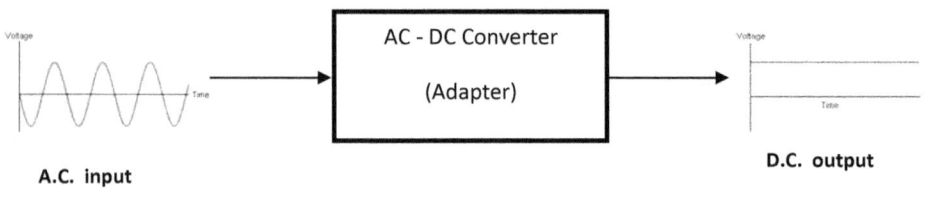

SAT Physics (Physics made simple)

Series VS Parallel Connections

Series connection means that the same amount of current flows in the circuit

Concepts:

1. Current is same every point [$I_T = I_1 = I_2$]

2. Voltage may be different on the bulbs but the sum should be same [$V_T = V_1 + V_2$]

3. Total resistance are normally high [$R_T = R_1 + R_2$]

4. If one bulb break the whole circuits fail [open switch at that bulb]

5. Total power supplied by cells equal sum of power used by bulb [$P_T = P_1 + P_2$]

Example 13.3

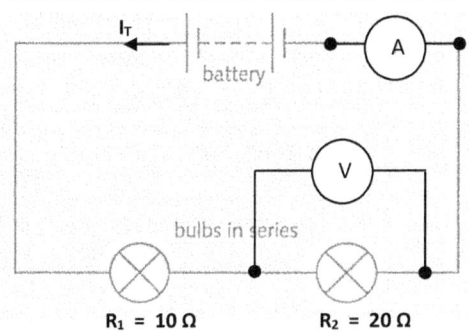

What are the readings on voltmeter and ammeter? Given that e.m.f. of the cell is 6 volt.

Solution: We will find "I_T" first

$R_T = R_1 + R_2$ → $R_T = 10 + 20 = 30\ \Omega$

$V_T = I_T R_T$ → $6 = I_T\ 30$ → $I_T = 0.2$ A

Since this is series connection, **ammeter reading will read 0.2 A**

To find reading on voltmeter we will use $V_2 = I_2 R_2$ → $I_2 = I_T = 0.2$ A

plug in $V_2 = 0.2 \times 20 = 4$ V

Therefore the reading on **voltmeter will be 4 V**

SAT Physics (Physics made simple)

Parallel connection means that the same amount of voltage are delivered on each node or appliance

Concepts:

1. Voltage is same one every node [$V_T = V_1 = V_2$]
2. Current are divided out to each device at the node [$I_T = I_1 + I_2$]
3. Total resistance are quite low [$1/R_T = 1/R_1 + 1/R_2$]
4. If one bulb break the others still work
5. Total power supplied by cells equal sum of power used by bulb [$P_T = P_1 + P_2$]

Example 13.4

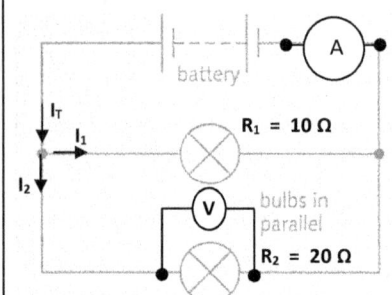

What are the currents thru each resistor ? What reading does ammeter shows ? Find also R_T (Given that e.m.f. of the cell is 60 volt)

Solution: Since we know that in parallel connection voltage are same

$V_T = V_1 = V_2 = 60$ V

We apply Ohm's law at each node

$V_1 = I_1 R_1$ → $60 = I_1 \cdot 10$ → $I_1 = 6$ A

$V_2 = I_2 R_2$ → $60 = I_2 \cdot 20$ → $I_2 = 3$ A

To find reading on ammeter we will use : $I_T = I_1 + I_2$

plug in $I_T = I_1 + I_2 = 6 + 3 =$ __9 A__

Therefore the reading on **ammeter will be 9 A**

$V_T = I_T R_T$ → $60 = 9 R_T$ → $R_T =$ __6.7 Ω__

SAT Physics (Physics made simple)

Other components

Diode is a semi-conductor made from germanium or silicon, normally used in electronic circuits.

Actual diode

Circuit symbol

Concepts of diode:

1. Diode has two poles + (anode) and - (cathode)
2. P-type for anode and N-type for cathode
3. Operating voltage of a diode is 0.7 V, then only it will function

Diode with d.c.

When a **diode is connected with d.c.** power supply it will act like a switch to prevent current from flowing in a wrong direction.

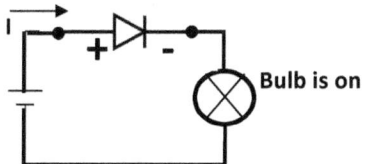

The arrow of showing the flow of current and diode are pointing in the same direction so diode will behave like a closed switch, allowing current to pass thru.

This is known as 'forward bias' connection.

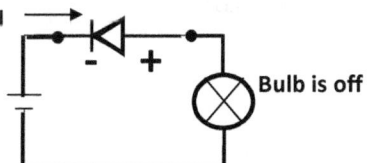

The arrow of showing the flow of current and diode are pointing in the opposite direction so diode will behave like an open switch, not allowing current to pass thru.

This is known as 'reverse bias' connection.

Diode with a.c.

When a **diode is connected with a.c.** power supply it will act like a rectifier (cutter) of current.

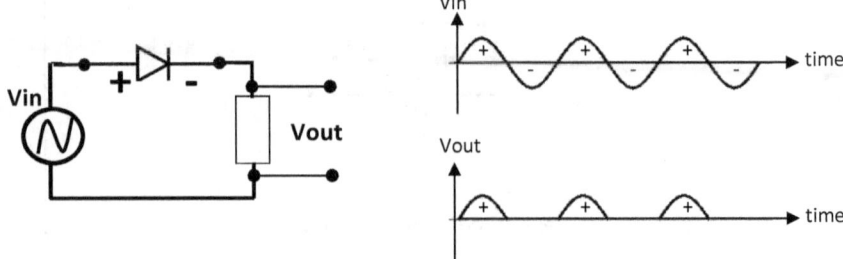

For the connection above the voltage meets the positive junction of diode first, therefore only positive voltage/current will be allowed to flow out. This is known as **'forward biased'** connection.

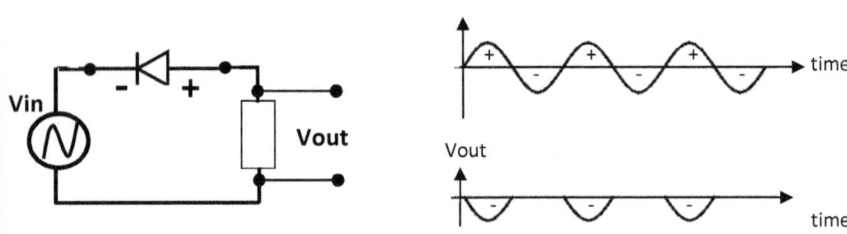

For the connection above the voltage meets the negative junction of diode first, therefore only negative voltage/current will be allowed to flow out. This is known as **'reverse biased'** connection.

Both circuits are considered to be a half-wave rectifier circuit, which doesn't give a complete d.c. voltage yet. We still need more diodes and capacitors to help filter out the desire voltage.

Capacitors

A **capacitor** is a device for storing charge, made up of two parallel plates with a space between them. The plates have an equal and opposite charge on them, creating a potential difference between the plates.

Inside a capacitor

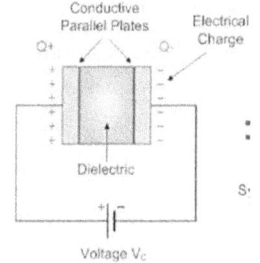

Actual capacitor

Circuit symbol

We can calculated total charged a capacitor can stored by using formula:

Charges= capacitance x voltage
$$Q = CV$$
(Coulomb)   (Farad) (Volt)

Remember that capacitance of a capacitor also depend on the distance between the plate and the area between the plates. So we can say that capacitance(C) is inversely proportional to distance(d) and capacitance is directly proportional to area(A) between the plates.

$$C \propto \frac{A}{d}$$

Capacitor in Series

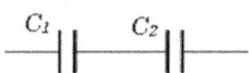

To find total capacitance

$$1/C_T = 1/C_1 + 1/C_2$$

Capacitor in Parallel

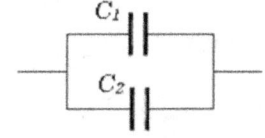

To find total capacitance

$$C_T = C_1 + C_2$$

SAT Physics (Physics made simple)

If a charge moves through a capacitor then the capacitor is said to have a **stored energy (U)**.

$$\text{Energy} = \text{charge} \times \text{voltage}$$
$$U = \frac{QV}{2}$$
$$\text{(Joules)} \quad \text{(Coulomb) (Volt)}$$

Example 13.5

A 10 volt e.m.f. battery is connected in series with a 50 Ω and an uncharged capacitor of 30 μF. a) What is the current through when the switch is closed after t = 0 second ?
b) What is current after a long time and how much charge is stored in a capacitor ?

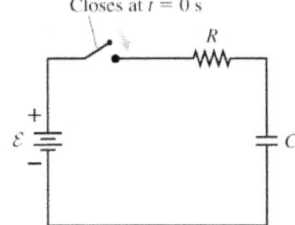

solution: a) we can apply Ohm's law at t = 0 sec
(ignore capacitor for now)

$$V = IR$$
$$10 = I \times 50$$
$$I = \underline{0.2 \text{ A}}$$

b) after a long time the charges will all be stored in a capacitor, so the current in the circuit will be equal to **zero**.

We can calculate charges by using $Q = C \times V$

$$Q = 30 \times 10^{-6} \times 10$$
$$Q = \underline{0.3 \text{ mC}}$$

Transistors

Transistor is a 3 terminal electronic device made from semiconductor material. Transistors have many uses such as amplifying signals, switching, voltage regulator, and modulation of signals.

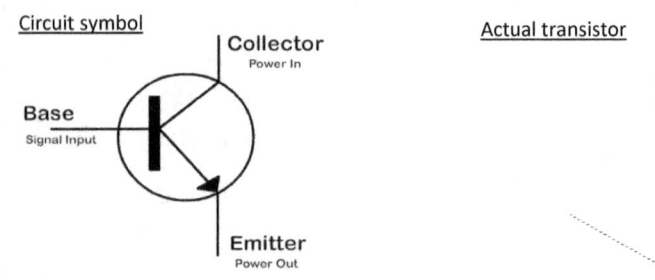

Practice
Chapter 13

Question 1 - 4

Given that battery has an e.m.f. of 40 Volt, resistor R_1 has 10 Ω, resistor R_2 has 30 Ω and resistor R_3 has 40 Ω of resistance.

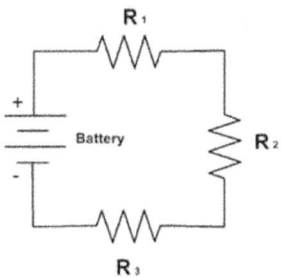

1. What is the total resistance of the circuit ?

    a) 5.5   Ω

    b) 6.3   Ω

    c) 10   Ω

    d) 40   Ω

    e) 80   Ω

2. What is the current through resistor R_3 ?

    a) 0.5   A

    b) 1.0   A

    c) 1.5   A

    d) 30   A

    e) 40   A

3. What is the potential difference across R_2 ?

   a) 0   V
   b) 10   V
   c) 15   V
   d) 20   V
   e) 40   V

4. What is the potential difference across R_1 given that R_3 burned out ?

   a) 0   V
   b) 10   V
   c) 15   V
   d) 20   V
   e) 40   V

Question 5 - 8

Given that battery has an e.m.f. of 40 Volt , resistor R_1 has 10 Ω , resistor R_2 has 20 Ω and resistor R_3 has 40 Ω of resistance.

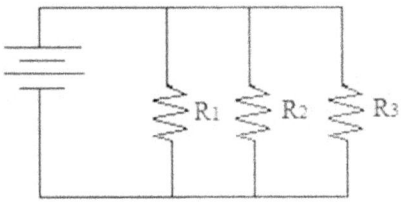

5. What is the total resistance of the circuit ?

   a) 5.7   Ω
   b) 6.3   Ω
   c) 10   Ω
   d) 40   Ω
   e) 70   Ω

SAT Physics (Physics made simple)

6. What is the current through resistor R_3 ?

   a) 0.5 A

   b) 1.0 A

   c) 1.5 A

   d) 30 A

   e) 40 A

7. What is the potential difference across R_2 ?

   a) 0 V

   b) 10 V

   c) 15 V

   d) 20 V

   e) 40 V

8. What is the total power supply by the battery?

   a) 40 W

   b) 100 W

   c) 150 W

   d) 280 W

   e) 400 W

9. What is the p.d. across the 5 Ω lamp ?

   a) 10 V

   b) 9.7 V

   c) 9.3 V

   d) 5 V

   e) 0.7 V

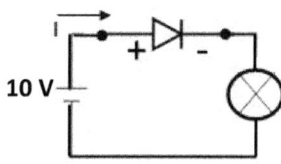

10. Which of the graph represents the graph of ' Voltage vs Current ' of a filament bulb ?

a)

b)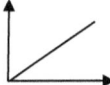

c)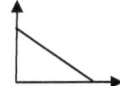

d)

e)

11. What is the total capacitance of the circuit below? Given that each capacitor has a capacitance of 20 µF

a) 10   µF
b) 20   µF
c) 30   µF
d) 40   µF
e) 80   µF

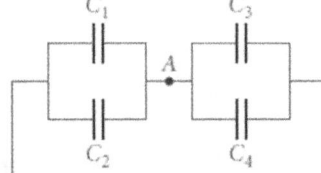

Scan me

SAT Physics (Physics made simple)

Answer: Chapter 13

Question	Answer	Explanation
1.	E	To find R_{total} in series we use: $R_T = R_1 + R_2 + R_3$ $R_T = 10 + 30 + 40 = 80\ \Omega$
2.	A	We use Ohm's Law here: $V_T = I_T R_T$ $40 = I_T \times 80$ $I_T = 0.5\ A$ $I_3 = I_T = 0.5\ A$
3.	C	Since this is a series circuit we know that $I_T = I_1 = I_2 = I_3$ $V_2 = I_2 \times R_2$ $V_2 = 0.5 \times 30 = 15\ A$
4.	A	In series, if one of the resistor is broken then it will act like an open switch so no current will be flowing out from the cell. Therefore current is **zero**.
5.	A	To find R_{total} in parallel we use: $1/R_T = 1/R_1 + 1/R_2 + 1/R_3$ $1/R_T = 1/10 + 1/20 + 1/40$ $1/R_T = 7/40$ $R_T = 40/7 \approx 5.7\ \Omega$
6.	B	Since this is a parallel circuit we know that $V_T = V_1 = V_2 = V_3$ $V_3 = I_3 \times R_3$ $40 = I_3 \times 40$ $I_3 = 1\ A$
7.	E	Since this is a parallel circuit we know that $V_T = V_1 = V_2 = V_3 = 40\ V$

Question	Answer	Explanation
8.	D	$I_1 = V_1 / R_1 = 40/10 = 4$ A $I_2 = V_2 / R_2 = 40/20 = 2$ A $I_3 = V_3 / R_3 = 40/40 = 1$ A $I_T = I_1 + I_2 + I_3 = 4 + 2 + 1 = 7$ A $\quad\quad\quad\quad P_T = I_T V_T$ $\quad\quad\quad\quad P_T = 7 \times 40$ $\quad\quad\quad\quad P_T = 280$ W
9.	C	This circuit is a series circuit so we use: $V_T = V_{DIODE} + V_{RESISTOR}$ $\quad\quad\quad\quad\quad\quad\quad\quad\quad\quad\quad\quad\quad 10 = 0.7 + V_{RESISTOR}$ $\quad\quad\quad\quad\quad\quad\quad\quad\quad\quad V_{RESISTOR} = 10 - 0.7 = 9.3$ V
10.	E	Graph of ' Voltage vs Current ' of a filament bulb will not be linear graph since heat will also be produced inside the bulb
11.	B	This question is about 2 sets of parallel capacitors connected in series $C_X = C_1 + C_2$ $\quad\quad C_Y = C_3 + C_4$ $C_X = 20 + 20$ $\quad\quad C_Y = 20 + 20$ $C_X = 40$ μF $\quad\quad\quad C_Y = 40$ μF To find C_{total} we use $1/C_{TOTAL} = 1/C_X + 1/C_Y$ $\quad\quad\quad\quad\quad\quad\quad\quad\quad\quad 1/C_{TOTAL} = 1/40 + 1/40$ $\quad\quad\quad\quad\quad\quad\quad\quad\quad\quad 1/C_{TOTAL} = 2/40$ $\quad\quad\quad\quad\quad\quad\quad\quad\quad\quad\quad C_{TOTAL} = 20$ μF

Electromagnetic Induction 14

Check-list:

- ✓ Magnetic field
- ✓ Magnetic force and flux
- ✓ Faraday's law
- ✓ Lenz's law
- ✓ Fleming left-hand rule
- ✓ Generator and Transformer

What is a magnet ?

- an object that attracts other metal

- an object that attract or repel other magnet

- an object that has two poles

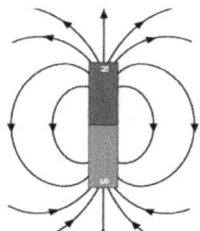

A **magnet** has a magnetic domain that are aligned perfectly from south to north. A **magnetic field** always flows from north to south. Strength of a magnet is defined by its number field lines.

Same pole of magnet would repel each other

Different pole would attract each other

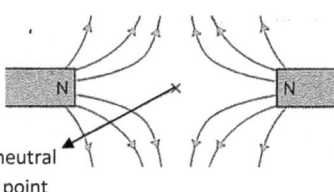

neutral point

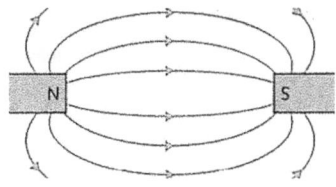

SAT Physics (Physics made simple)

Current carrying wire (Right-hand grip rule)

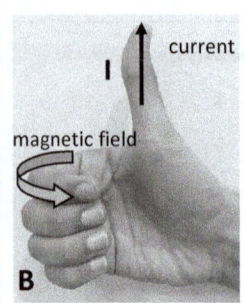

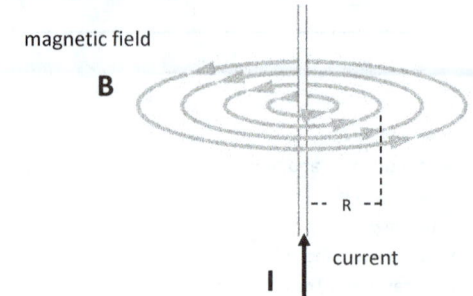

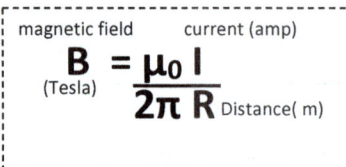

$$B_{(Tesla)} = \frac{\mu_0 \, I}{2\pi \, R}$$

where μ_0 (permeability of free space) = $4\pi \times 10^{-7}$ Tm/A

Magnetic field is strongest when it is near the wire (current) and weakest when it is further.

Current coming out of the page **Current going into the page**

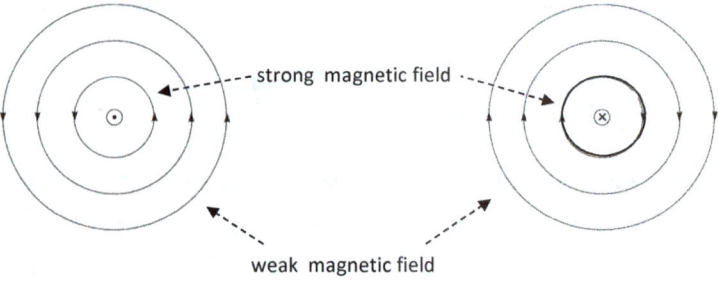

SAT Physics (Physics made simple)

Force between two current carrying wire can be determined using the concept of right hand rule.

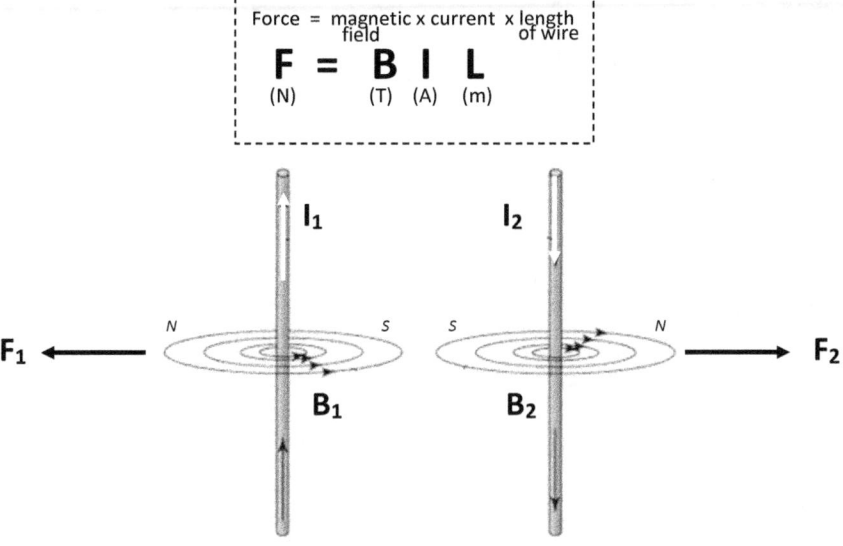

Since wire 1 is carrying current upward and wire 2 carrying current downward the forces due to magnetic field between the two wire will be a repulsive force in this case.

Solenoid

Solenoid can be used to demonstrate the right hand rule and the concept of electromagnet

Example 14.1

In a wire of length 2 m has a current of 5 A flowing through at the distance of 2 cm from the wire calculate:

a) the magnetic field   b) the force experience by a charge

Solution: a) We use $B = \dfrac{\mu_0 I}{2\pi R}$ we have the $I = 5A$, $R = 2 \times 10^{-2}$ m

$$B = \dfrac{4\pi \times 10^{-7} \times 5}{2\pi \times 0.02}$$

$$B = 5 \times 10^{-5} \text{ T}$$

b) To find the force we use $F = BIL$

$F = 5 \times 10^{-5} \times 5 \times 2$

$F = \underline{5 \times 10^{-4} \text{ N}}$

Magnetic force on a moving charge

If a charge moves through a region of magnetic field a force is experienced by the charge.

These force can be calculated by looking at them into three cases:

1. A charge moving at 90° or perpendicular to the magnetic field

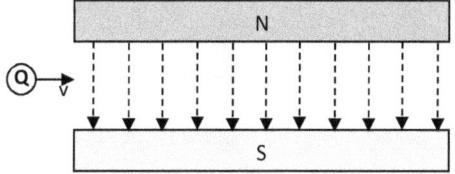

$$F_{(N)} = Q_{(C)} \, v_{(m/s)} \, B_{(T)}$$

2. A charge moving at angle θ

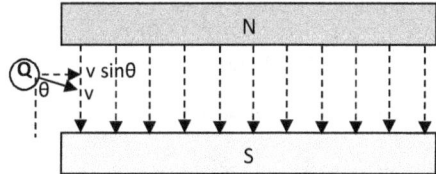

$$F_{(N)} = Q_{(C)} \, v\sin\theta_{(m/s)} \, B_{(T)}$$

3. A charge moving parallel to the magnetic field

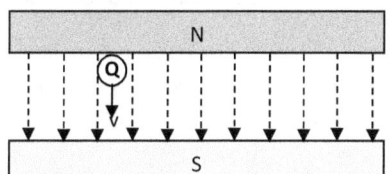

$$F_{(N)} = 0$$

Magnetic flux (Φ)

Magnetic flux is the amount of magnetic field strength(B) passing through a given area (A)

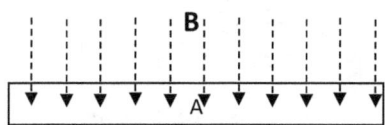

$$\Phi_{(Wb)} = B_{(T)} \cdot A_{(m^2)}$$

Faraday's law

Faraday discovered that if a coil of wire cuts through the magnetic field then a current will be induced. In other word, changing in magnetic flux($\Delta\Phi$) causes an induced e.m.f.(ε) on the wire.

$$\varepsilon_{(v)} = \frac{\Delta\Phi_{(wb)}}{\Delta t_{(s)}}$$

Lenz's law

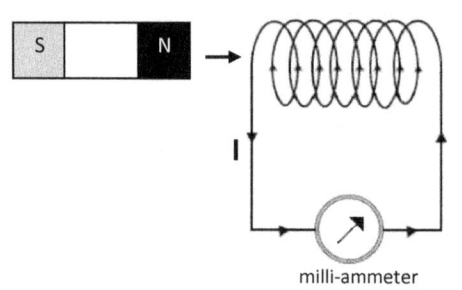

To increased the induced current we can :

i) move the magnet faster

ii) use stronger magnet

iii) increased number of turns in coil

iv) increased number of flux

*reversing the direction of magnet will reverse the direction of induced current

**if a magnet or a coil is not moving then no current is induced (no field being cut)

Fleming left hand rule

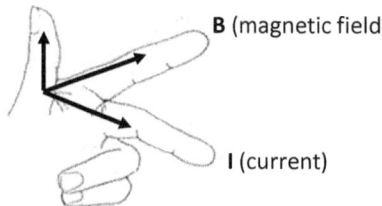

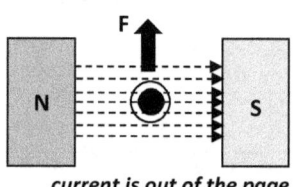

current is out of the page

According Fleming, if there is a current on a wire running through magnetic field then a force will be induced. Or if there is a force on a wire cutting through magnetic field then the current will be induced.

SAT Physics (Physics made simple)

If we reverse the direction of the current without changing the direction of the magnetic field then force will also reverse its direction.

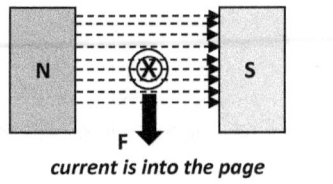

current is into the page

We can use the concept of Fleming rule to help us determine the direction of the spin of a simple d.c. motor.

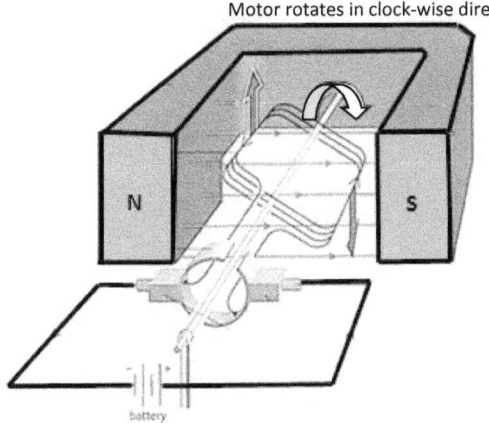

Motor rotates in clock-wise direction

If we want the motor to spin faster:

1) Use strong magnet

2) Increase e.m.f. of the cell

3) Increase number of coils

Fleming left-hand rule can be apply also with the movement of electrons through a magnetic field region by shifting to right hand

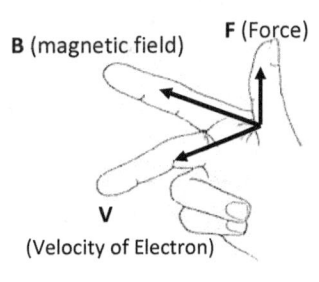
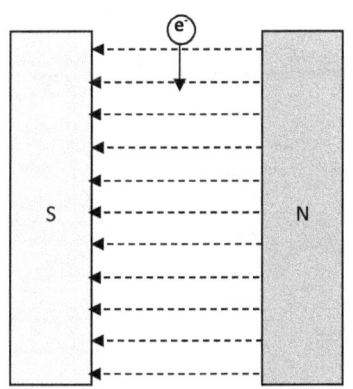

Applying the rule here electron will move out of the page curved up towards the south pole of the magnet

Generator

Generator uses the principal of electromagnetic induction to generate electricity by cutting through the magnetic field. An output of this generator will be alternating current, the size depend on the spinning rate and the strength of the magnetic field.

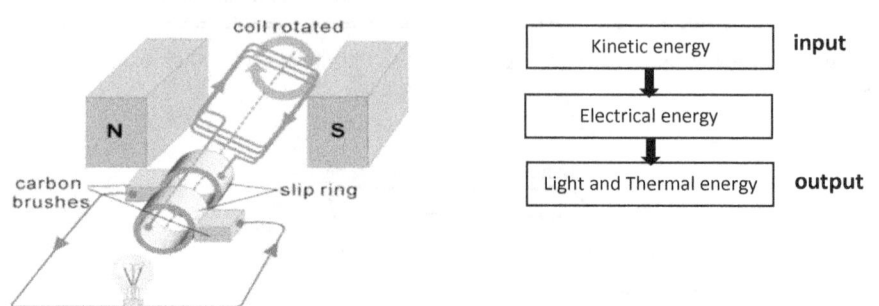

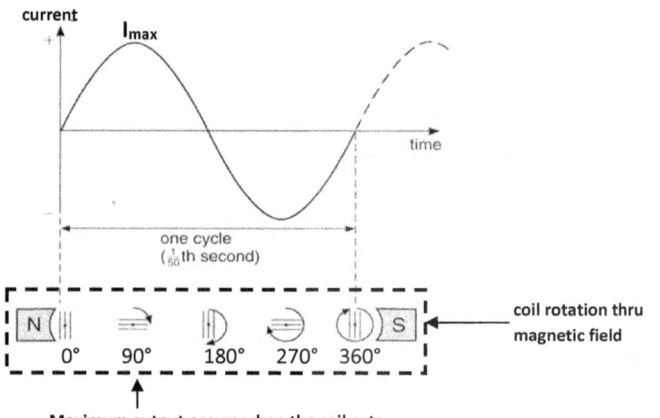

Maximum output occurs when the coil cuts the magnetic field at 90°

$Power_{average} = I_{rms} \times V_{rms}$

$I_{rms} = \dfrac{I_{max}}{\sqrt{2}}$

$V_{rms} = \dfrac{V_{max}}{\sqrt{2}}$

$P_{avg} = \dfrac{I_{max} \times V_{max}}{2}$

Transformer

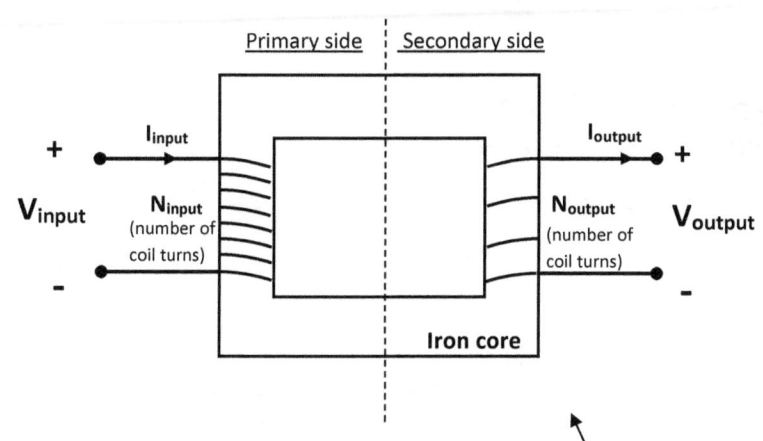

A transformer is a device used to increased(step-up) or decreased(step-down) the output voltage for transmitting electricity over a long distance. A transformer only woks with alternating current (a.c.) because there is a change in magnetic field.

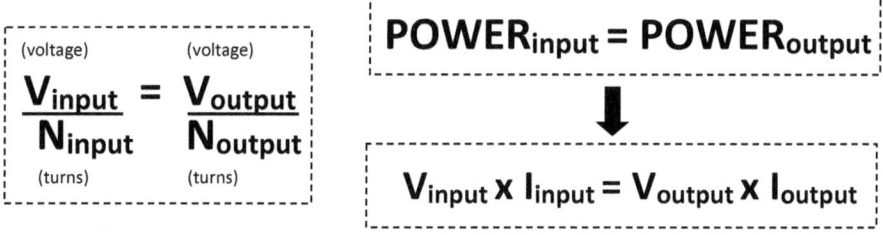

$$\frac{V_{input}}{N_{input}} = \frac{V_{output}}{N_{output}}$$
(voltage/turns)

$$POWER_{input} = POWER_{output}$$

$$V_{input} \times I_{input} = V_{output} \times I_{output}$$

How does a transformer works ?

1. a.c. flows through the primary coil this causes alternating magnetic field in the iron core

2. secondary side's coil cuts the alternating magnetic field which according to Lenz's law a voltage(a.c.) will be induced on the secondary side

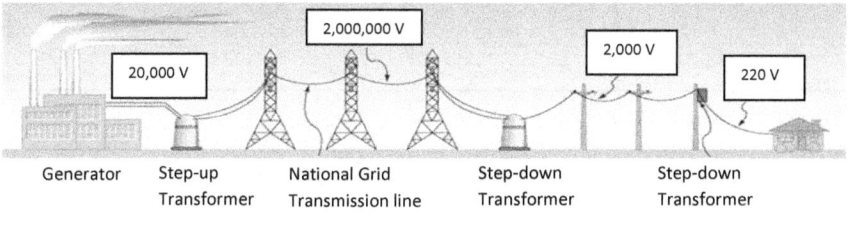

Why do we have to increase the voltage before transmitting over the transmission line?

Because we want the reduce the current by increasing the voltage, thus this will help reduce the heat produced on the wire (minimizing heat loss). Current is inversely proportional to voltage.

$$\text{Heat loss (joules)} = I^2 \times R \times t$$
$$\text{(amp)} \quad \text{(ohm)} \quad \text{(second)}$$

Example 14.2

A step-up transformer has an input voltage of 440 V and the coil are winded 300 turns while the secondary side is winded 6000 turns and has an output current of 0.5 A, given that the transformer is 100% efficient calculate

      a) output voltage   b) current supply to the transformer

Solution: a) We use $V_{in} / N_{in} = V_{out} / N_{out}$

                              $440/300 = V_{out} / 6000$

                              $V_{out} = \underline{8,800 \text{ V}}$

b) To find the current input we use $P_{in} = P_{out}$

                              $I_{in} \times V_{in} = I_{out} \times V_{out}$

                              $I_{in} \times 440 = 0.5 \times 8800$

                                   $I_{in} = \underline{10 \text{ A}}$

Practice
Chapter 14

1. What is the magnetic field through a square coil of length 5 cm if there is a flux linkage of 3 Wb?

    a)   0.0075 T

    b)   0.0015 T

    c)   15   T

    d)   750   T

    e)   1200 T

2. If the coil was moved around the magnet rapidly once, which of the following is TRUE ?

    a)   ammeter does not move

    b)   ammeter move rapidly and stop

    c)   the coil become magnetized

    d)   the coil produced a spark

    e)   a magnetic force pushed the coil way

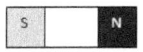
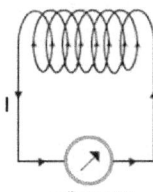
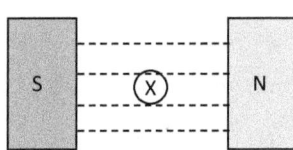

3. Which direction is the force on the wire that carries current into the page ?

    a)   upward

    b)   downward

    c)   eastward

    d)   westward

    e)   out of page

4. Which direction is the force acting on electron that is moving into the page ?

   a)   upward
   b)   downward
   c)   eastward
   d)   westward
   e)   out of page

Question 5 - 7

A 10 cm wire carries a current of 8 A is placed between the magnetic field of 5 T

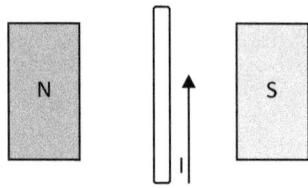

5. How much force is experienced by the wire?

   a)   4   N
   b)   6   N
   c)   10   N
   d)   12   N
   e)   40   N

6. Which direction is the force acting on the wire ?

   a)   upward
   b)   downward
   c)   eastward
   d)   into the page
   e)   out of page

7. If the magnetic field is reduced by half and the force is remain constant, then how much current must be applied ?

   a) 4   A
   b) 6   A
   c) 8   A
   d) 16   A
   e) 20   A

8. Which of the following is TRUE about electromagnetic induction?

   a) if a force is applied to a wire then a magnetic field is created
   b) if a current carrying wire pass through a magnetic field then no force is created
   c) if a current carrying wire pass through a magnetic field then a force is created
   d) if a magnetic field is created then a force is applied to the wire
   e) if a electrons flows in a wire then a force is applied to the wire

9. How much e.m.f. is induced if we cut a wire pass through magnetic flux of 20 Wb in 0.5 seconds ?

   a) 40   V
   b) 20   V
   c) 10   V
   d) 5   V
   e) 2   V

10. What is the output current of a transformer that has an a.c. input of 6 A and a primary coil winding of 120 turns to the secondary coil winding of 3600 turns ?

a) 0 A
b) 0.2 A
c) 0.3 A
d) 0.5 A
e) 0.6 A

Scan me

11. What is the output current of a transformer that has a d.c. input of 0.1 A and a primary coil winding of 120 turns to the secondary coil winding of 3600 turns ?

a) 0 A
b) 0.2 A
c) 0.3 A
d) 0.5 A
e) 0.6 A

SAT Physics (Physics made simple)

Answer: Chapter 14

Question	Answer	Explanation
1.	E	To find magnetic field we use $B = \dfrac{\Phi}{A} = \dfrac{3 \text{ wb}}{25 \times 10^{-4} \text{ m}^2} = 1200$ T
2.	B	By moving the coil around the magnet, we are still cutting through the magnetic field, an e.m.f. will be induced for when there is a movement. Then it will stop when the movement stop.
3.	A	We apply Fleming left hand-rule, we get a force pointing upward
4.	B	We alter the left hand rule by using right-hand instead since electrons and current flow in the opposite direction, we get a force pointing downward
5.	A	To find Force we use $F = B \times i \times L$ $F = 5 \times 8 \times 0.1$ $F = 4$ N

Question	Answer	Explanation
6.	D	Applying Fleming's left-hand rule, we get a force pointing into the page
7.	D	To find current we apply $F = B \times i \times L$ $4 = 2.5 \times i \times 0.1$ $i = 16$ A
8.	C	Electromagnetic induction means that there is a force, a magnetic field and a current. If any of the two are present then the third will show up. In this case we are saying that a current is moving through a magnetic field then a force will be experienced by the wire.
9.	A	We use the formula $$\varepsilon = \frac{\Delta \Phi}{\Delta T} = \frac{20 \text{ wb}}{0.5 \text{ s}} = 40 \text{ V}$$
10.	B	Transformer with an a.c. input $I_{in} \times N_{in} = I_{out} \times N_{out}$ $6 \times 120 = I_{out} \times 3600$ $I_{out} = 0.2$ A
11.	A	Transformer with a d.c. input will not give any output current (or voltage), we can say that output current is 0 A.

Quantum Physics   15

Check-list:

- ✓ Inside atom
- ✓ Rutherford model of atom
- ✓ Photoelectric effect & Bohr's model of atom
- ✓ Half-life and Radioactivity

What is an atom ?

Atoms are building blocks of matter that are composed of protons, neutrons and electrons. An element is normally defined according to number of protons, neutrons and electrons inside it. Let's take a look inside a helium atom.

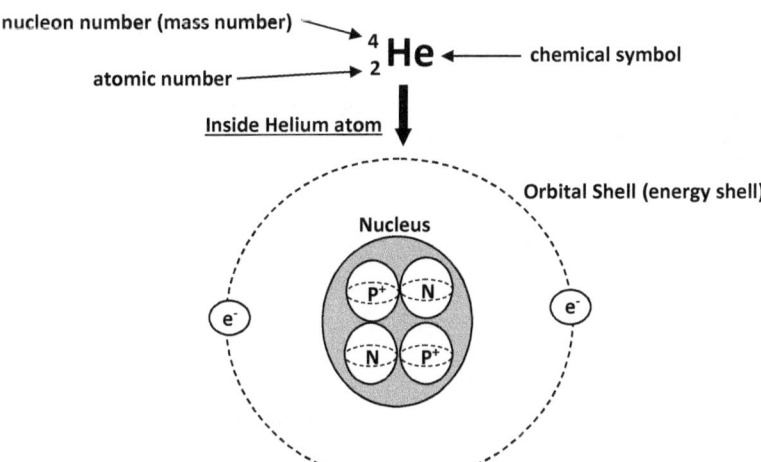

nucleon number (mass number) — $^4_2$He ← chemical symbol

atomic number

Inside Helium atom

Concept of atom:

i) Proton have a positive charge, electron has a negative charge and neutron has no charge

ii) Number of protons = Number of electrons

iii) Nucleon number = number of proton + number of neutron

iv) Atomic number = number of proton

v) Electrons are lightest (almost massless)

Matter vs Antimatter

Atom of Matter	**Atom of Antimatter**
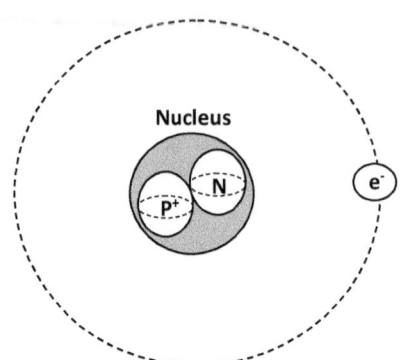	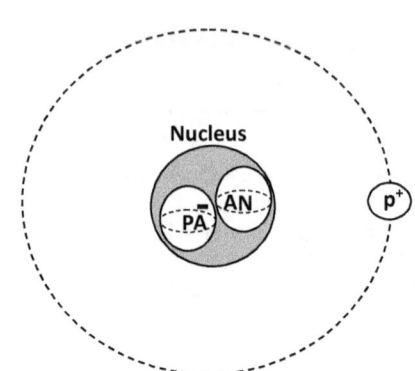

Matters and antimatter are identical in **mass** but opposite in charge. When matter and antimatter collide a burst of gamma ray is created (energy is released).

Electron	Positron
Proton	Antiproton
	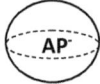
Neutron	Antineutron

SAT Physics (Physics made simple)

Rutherford model of atom

Ernest Rutherford perform an experiment by shooting alpha particles (Helium atoms) into a thin gold foil. He discovered that most of the alpha particles are able to pass thru easily, however some deflected and some bounced back.

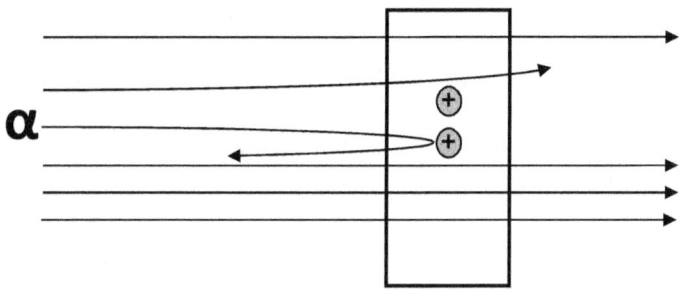

His conclusions were:

(i) Atoms are MOSTLY EMPTY space

(ii) All atom's mass are concentrated at the tiny part called nucleus

(iii) Nucleus is considered POSITIVELY CHARGED because of protons

(iv) Electrons orbit nucleus just like how planets orbit the Sun

Photoelectric effect

When a high-frequency electromagnetic wave hits a metal and an electron is emitted off this is known as **photoelectric effect**, while the emitted electrons are called **photoelectron**. According to Einstein quantum theory, which state that light a bundle of energy (vibration of photons) so energy is transmitted to a body. Max Planck then explained that the concept of quantum theory of light by stating that light comes in a packet of energy have both particles and wave properties. Each quantum is related to definite amount of energy which can be calculated by formula below.

Energy = Planck constant x frequency

$$E = h f$$
(Joules)   (Hertz)

h → Planck constant

$h = 6.63 \times 10^{-34}$ J·s

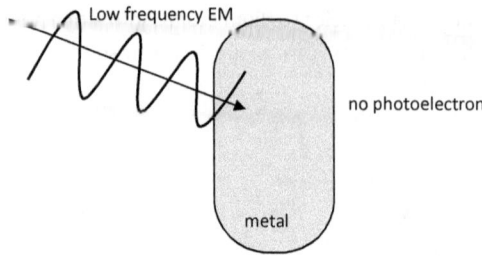
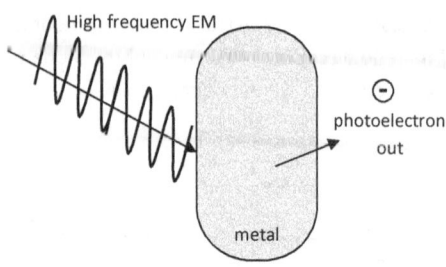

No emission of electron occur because the frequency of electromagnetic wave is lower than threshold frequency (f_0).

Emission of electron occur because the frequency of electromagnetic wave is geater than threshold frequency (f_0).

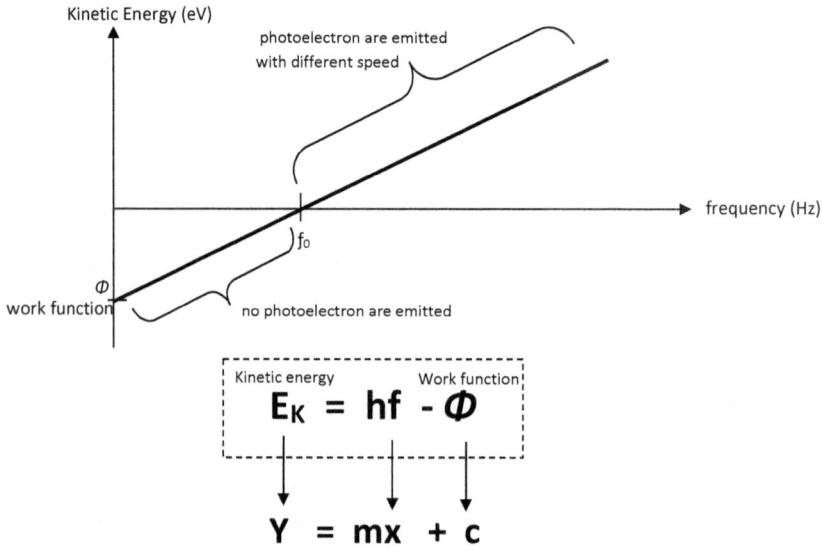

$$E_K = hf - \Phi$$

$$Y = mx + c$$

Concept of photoelectric effect:

1. Work function(Φ) is the minimum amount of energy needed to push photoelectron out from the metal. Work function is the Y-intercept(c) of the line.

2. Slope(m) is equivalent to Planck constant (h).

3. The intensity of light is directly proportional to amount of electron emitted out.

4. The intensity of light has nothing to do with the speed of emitted photoelectron.

5. Any frequency lower than threshold frequency (f_0) will not allow any emission of photoelectron.

6. Each type of metal have a different threshold frequency (f_0) and a different speed of emitted photoelectron.

7. 1eV is amount of energy an electron have when 1 volt is applied to it. (1eV = 1.6×10^{-19} J).

Example 15.1

An electromagnetic wave of wavelength 350 nm falls on a certain metal with a work function of 2.5×10^{-19} J. Calculate:

a) amount of energy carried by photon before impact   b) speed of photoelectron

Solution: a)   We use $E = hf$ we know $\lambda = 350 \times 10^{-9}$ m, $V_c = 3 \times 10^8$ m/s

$E = h(V_c/\lambda)$

$E = (6.63 \times 10^{-34})[(3 \times 10^8)/(350 \times 10^{-9})]$

$E = \underline{5.68 \times 10^{-19}}$ J

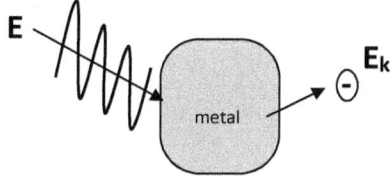

b)   To find the speed of photoelectron

$E_k = hf - \Phi$

$E_k = 5.68 \times 10^{-19} - 2.5 \times 10^{-19}$

$E_k = 3.18 \times 10^{-19}$ J

$\frac{1}{2} m_{electron} v^2 = 3.18 \times 10^{-19}$ J

$v^2 = \dfrac{2 \times 3.18 \times 10^{-19}}{9.1 \times 10^{-31}}$

$v = \underline{8.36 \times 10^5}$ m/s

Neil's Bohr model of atom

Bohr's model of atom explained why electrons orbit around the nucleus without falling into it. Each electrons orbit in specific energy shell (levels), the most stable orbit is closest to the nucleus. Bohr's made an observation on hydrogen atom since it consists only of one proton and one electron.

Concept of Bohr's model:

1) Electron orbiting in lowest energy level is said to be in ground state (n = 1)

2) Electrons can be raised to higher energy level(excited) by giving a quantized amount of energy

3) Electrons can fall back down to ground state (if it was excited) by releasing the absorption energy

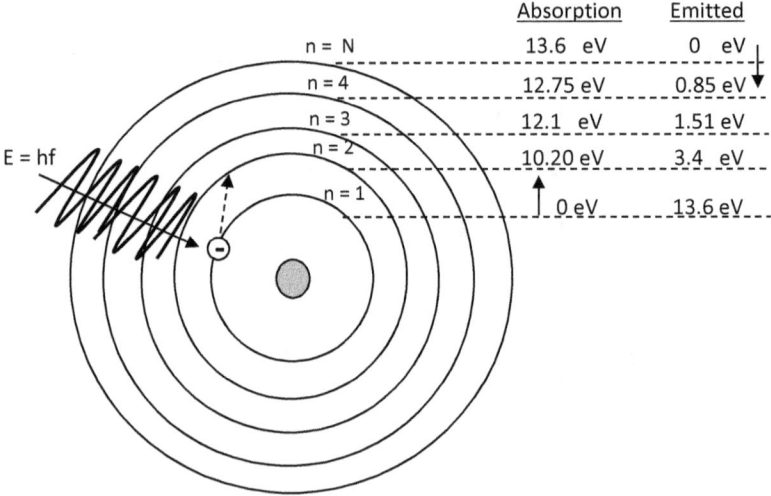

Energy level for Hydrogen atom

To find energy emitted or absorb by an electron we the formula:

$$E_{\text{(Joules)}} = \frac{h \; c \;\text{speed of light}}{\lambda \;\text{(meter)}}$$

Example 15.2

List all possible energies in eV of emitted photons from n = 4 of Hydrogen atom

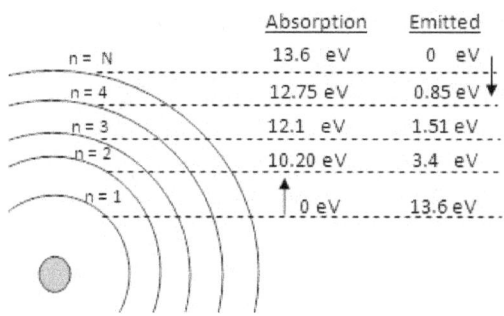

	Absorption	Emitted
n = N	13.6 eV	0 eV
n = 4	12.75 eV	0.85 eV
n = 3	12.1 eV	1.51 eV
n = 2	10.20 eV	3.4 eV
n = 1	0 eV	13.6 eV

Solution:

We can look at transition from $E_{n=4}$ to $E_{n=3}$ → 1.51 - 0.85 = 0.66 eV

$E_{n=4}$ to $E_{n=2}$ → 3.4 - 0.85 = 2.55 eV

$E_{n=4}$ to $E_{n=1}$ → 13.6 - 0.85 = 12.75 eV

$E_{n=3}$ to $E_{n=2}$ → 3.4 - 1.51 = 1.89 eV

$E_{n=3}$ to $E_{n=1}$ → 13.6 - 1.51 = 12.09 eV

$E_{n=2}$ to $E_{n=1}$ → 13.6 - 3.4 = 10.2 eV

Compton effect

If a very high energy photons hit a substance then scattering of a photon off an electrons could cause the electromagnetic wave and recoil electron to a conserved momentum.

$$p = \frac{h}{\lambda}$$

where p = momentum, h = Planck constant, λ = wavelength

When an electromagnetic wave falls on a substance we notice that we get two product, one is a new electromagnetic wave with a lower and an electron. Now we can apply law of conservation of momentum here by saying that $\Sigma P_i = \Sigma P_f$. According to Louis De Broglie, the emitted electron behave like a wave. This allow us to use concept of momentum to calculate the wavelength of electron by using formula for **Broglie wavelength (λ_B)**.

$$\Sigma P_i = \Sigma P_f$$
$$P = P' + P_{ELECTRON}$$
$$\frac{h}{\lambda} = \frac{h}{\lambda'} + \frac{h}{\lambda_B}$$

$$\lambda_B = \frac{h}{m_{electron} \times V_{electron}}$$

Half-life

Half-life of a substance is the time taken for a radioactive element's activity level become half of the original value, also known as radioactive decay. **Radioactive decay** is the process by which the unstable isotopes releases its energy in order to become more stable. **Isotopes** are elements that have same number of protons but different number of neutrons. According to Einstein's mass energy equivalency, mass(defect) is converted to energy.

Energy = mass × speed of light²

$$E = m c^2$$

(Joules) (Kg) (3×10^8 m/s)

Activity level is measured in Becquerel (Bq), 100 Bq is equal to 100 nuclei disintegration per second. A radioactive element will give out energy and then the rate of releasing will decrease according to exponential decay graph.

Graph of half-life

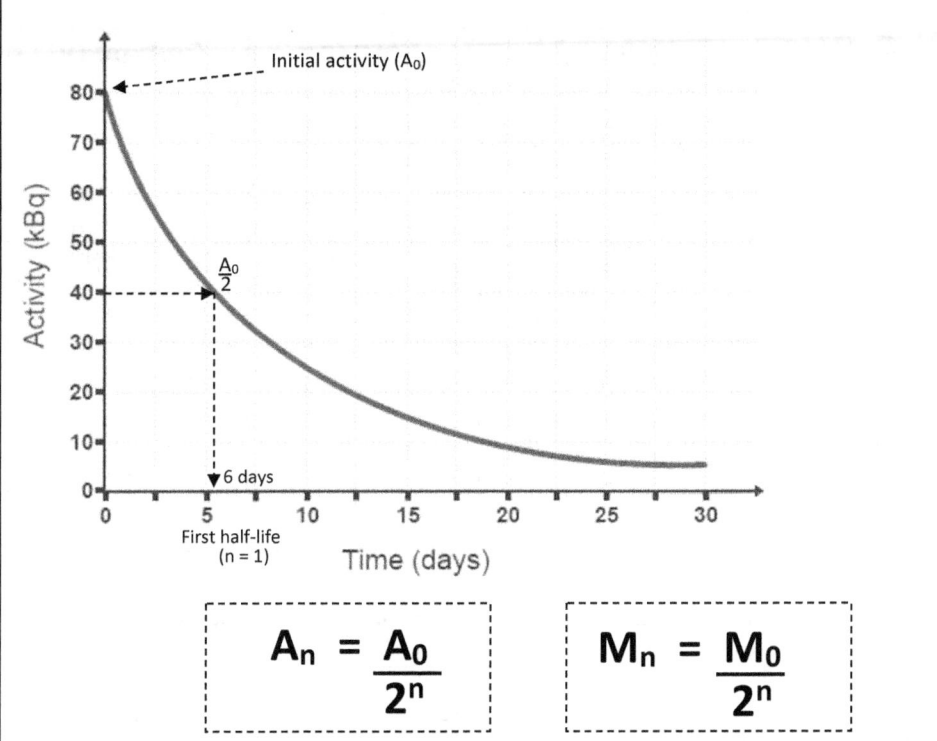

$$A_n = \frac{A_0}{2^n} \qquad M_n = \frac{M_0}{2^n}$$

where 'n' is the number of half-life

'A_n' is the activity left after n half-life and 'A_0' is the initial activity

'M_n' is the mass left after n half-life and 'M_0' is the initial mass

Example 15.3

A sample of uranium-235 has a mass of 240g and a half-life of 4 years. After 20 years how much mass is left over?

Solution: we first calculate how many half-life has pass (n) by using $n = 20 \div 4 = 5$ half-life

Now we can apply the formula for mass → $M_n = \dfrac{M_0}{2^n} = \dfrac{240}{2^5} =$ __7.5 grams left__

3 kinds of radiation

Types of radiation	Alpha (α)	Beta (β)	Gamma (γ)
	Composed of 2 protons and 2 neutrons	High speed electron	Electromagnetic wave
Relative Charge	+ 2	- 1	0
Mass	Identical to helium nucleus	Very small	No mass
Speed	0.1c	0.9c	c (speed of light)
Ionizing effect	Strong	Weak	Very little
Effect on magnetic field and electric field	Deflect	Deflect most	No Deflection

Penetration ability

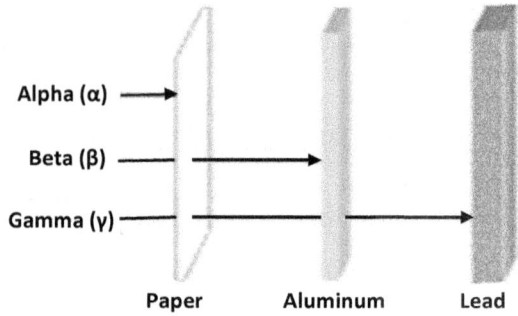

Alpha (α) — Paper
Beta (β) — Aluminum
Gamma (γ) — Lead

Deflection on field

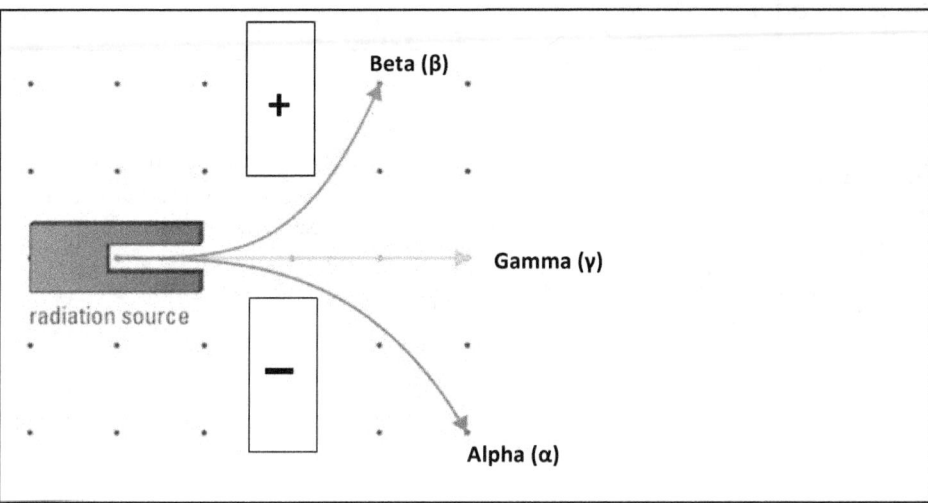

Decay equation

Alpha decay occurs when there is a release in *alpha particles* or *helium nucleus*.

$$^{238}_{92}U \rightarrow\, ^{234}_{90}Th\, +\, ^{4}_{2}He$$

Beta decay occurs when there is a release in *beta particles* or an *electron* and an *antineutrino*.

$$^{231}_{53}I \rightarrow\, ^{231}_{54}Xe\, +\, ^{0}_{-1}e\, +\, \bar{v}$$

Gamma decay occurs when there is a release of energy in form of electromagnetic wave.

$$^{238}_{92}U \rightarrow\, ^{238}_{92}U\, +\, \gamma$$

Nuclear Fusion and Fission

Nuclear fusion occurs when a very light nuclei joined together to form a heavier nuclei and a burst of energy.

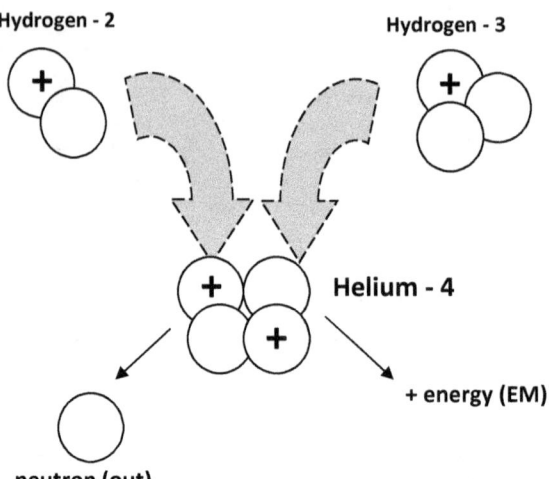

Hydrogen isotopes (H_2 and H_3) joined to form helium atom and electromagnetic radiation. This process is known to occur in core of the Sun at 15,000,000°C.

Nuclear fission occurs when a high speed neutron is shot into a uranium isotopes to trigger a chain reaction. This process normally occurs in a nuclear reactor.

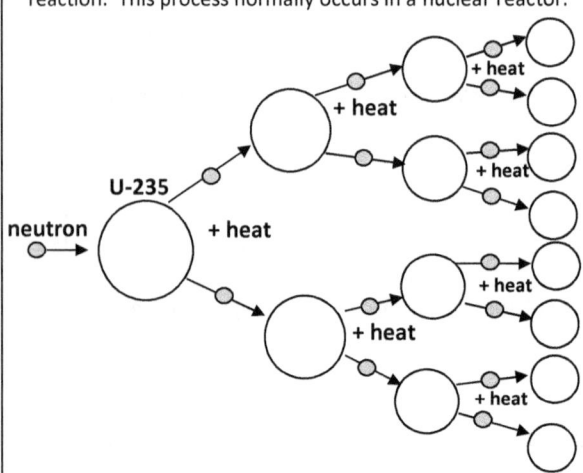

A **nuclear reactor** uses the process of controlled chain reaction and a release of thermal energy at a steady rate. This thermal energy is used to make steam for the turbine to spin in order to generate electricity.

Practice
Chapter 15

1. Which particles below is the heaviest and positively charged ?

    a) Proton

    b) Neutron

    c) Nucleus

    d) Electron

    e) Positron

2. Which of the following is TRUE about Rutherford model of atom ?

    a) atoms are not an empty space

    b) the entire mass of atom's are located in the nucleus

    c) nucleus is considered to be neutral

    d) electrons do not orbit nucleon

    e) atoms are positively charged

3. What are the missing values of 'x' and 'y' equal to ?

$$_{88}^{226}Ra \rightarrow \, _{y}^{x}Rn + \, _{2}^{4}He$$

   a) x = 230,   y = 90

   b) x = 230,   y = 86

   c) x = 222,   y = 90

   d) x = 222,   y = 88

   e) x = 222,   y = 86

4. According to question 3 what kind of decay took place ?

   a) Alpha

   b) Beta minus

   c) Beta plus

   d) Gamma

   e) Radium

5. How much energy is carried by photons with a frequency of 2.5×10^{17} Hz ?

   a) 1.66×10^{-15} J

   b) 1.66×10^{-16} J

   c) 2.66×10^{-15} J

   d) 2.66×10^{-16} J

   e) 3.66×10^{-15} J

6. What is the mass left over of a 500g radioactive substance that has a half-life of 40 days 120 days before ?

   Scan me

   a) 62.5 g

   b) 125 g

   c) 250 g

   d) 1000 g

   e) 4000 g

7. What is Broglie wavelength of an electron that moves with a speed of 9×10^{4} m/s ?

   a) 8.09 nm

   b) 7.32 nm

   c) 6.54 nm

   d) 3.25 nm

   e) 2.16 nm

Question 8 - 9 use the information below

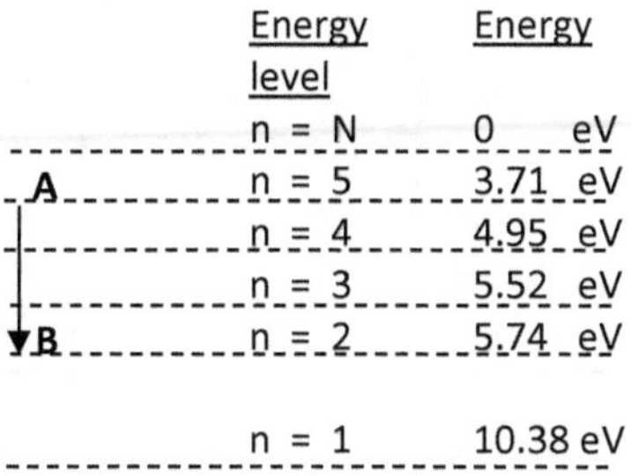

	Energy level	Energy
	n = N	0 eV
A	n = 5	3.71 eV
	n = 4	4.95 eV
	n = 3	5.52 eV
▼ B	n = 2	5.74 eV
	n = 1	10.38 eV

8. What is the energy released from A to B transition ?

   a) 8.09   eV

   b) 5.43   eV

   c) 3.00   eV

   d) 2.03   eV

   e) 0.79   eV

9. Which energy transition is NOT possible ?

   a) 8.09   eV

   b) 6.67   eV

   c) 5.43   eV

   d) 2.03   eV

   e) 0.79   eV

Answer: Chapter 15

Question	Answer	Explanation
1.	C	Nucleus contain protons and neutrons, therefore it is the heaviest and contain a positive charge
2.	B	According Rutherford model of atom " Almost the whole mass of an atom are concentrated at the nucleus, because electrons are very small "
3.	E	$$^{226}_{88}Ra \rightarrow\ ^{x}_{y}Rn + ^{4}_{2}He$$ $$226 = x + 4 \rightarrow x = 222$$ $$88 = y + 2 \rightarrow y = 86$$
4.	A	Alpha particles are similar to Helium nucleus, it contain 2 protons and 2 neutrons
5.	B	We use $E = h\ f$ $E = (6.63 \times 10^{-34}) \times (2.5 \times 10^{17})$ $E = 1.66 \times 10^{-16}$ J
6.	E	Mass before means we have to double it 500 g　　1000g　　2000g　　4000g earlier　0 day → 40 days → 80 days → 120 days So 120 days earlier we have 4000 grams
7.	A	To find Broglie wavelength we use 9×10^4 $$\lambda_B = \frac{h}{m_{electron} \times V_{electron}}$$ $$\lambda = \frac{6.63 \times 10^{-34}}{9.11 \times 10^{-31} \times (9 \times 10^4)}$$ $$\lambda = 8.09 \times 10^{-9} \text{ or } 8.09 \text{ nm}$$

Question	Answer	Explanation
8.	D	at A n = 5 → 3.71 eV at B n = 2 → 5.74 eV Difference between them = 5.74 - 3.71 = 2.03 eV
9.	A	a) 8.09 eV → NOT POSSIBLE b) 6.67 eV → n = 5 to n = 1 c) 5.43 eV → n = 4 to n = 1 d) 2.03 eV → n = 5 to n = 2 e) 0.79 eV → n = 4 to n = 2

Special Relativity 16

Check-list:

- ✓ Law of special relativity
- ✓ Relativity of time, length and mass
- ✓ Relativistic energy

Law of special relativity

Two postulates of Einstein law of special relativity:

1. All the laws of physics are the same in all inertial reference frames
2. The speed of light in vacuum is the same in all inertial frames ($V_c = 3 \times 10^8$ m/s) regardless of the motion of both the observer or the source

If a person is running with a speed of $0.4V_c$ on the space-craft that was moving at the speed of $0.7V_c$ then:

According to Newton's Law

$V_{relative} = 0.4V_c + 0.7V_c = 1.1\ V_c$

According to Einstein's Law

$V_{relative} = 0.4V_c + 0.7V_c \neq 1.1\ V_c$
$V_{relative} = 1\ V_c$

Relativistic factor:
$$Y = \frac{1}{\sqrt{1 - (v/c)^2}}$$

As we move faster (v) the value of the denominator of Y gets smaller nearly approaching zero.

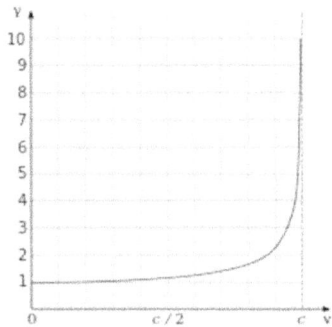

SAT Physics (Physics made simple)

Relativity of time, length and mass

As we move close to the speed of light the time would be moving slower, for instance if we let the person travels on space-craft in space at the speed of $0.9999V_c$ for 1 year (of space-craft time) then when he returned back according to Einstein the time on Earth would have been 71 years.

Time dilation can be calculated using:

$$T = t_0 \cdot Y = \frac{t_0}{\sqrt{1-(v/c)^2}}$$

v - speed that the object is moving

c - speed of light (3×10^8 m/s)

t_0 - time measured on moving object

T - time measured on Earth

As we move close to the speed of light the length would be getting smaller, for instance if a 100 meter long space-craft is moving with the speed of $0.5V_c$ a person on Earth would measures the length of the space-craft to be about 43.5 meter.

Length contraction can be calculated using:

$$L = L_0 \div Y = L_0 \cdot \sqrt{1-(v/c)^2}$$

v - speed that the object is moving

c - speed of light (3×10^8 m/s)

L_0 - original length of the object

L - length measured by stationary observer

As we move close to the speed of light the mass would seems to increase, for instance if a space-craft of mass 1000 kg is moving in space at the speed of $0.9V_c$ then it would appear that the space-craft is having a mass of 2300 kg.

mass effect can be calculated using:

$$m = m_0 \cdot Y = \frac{m_0}{\sqrt{1-(v/c)^2}}$$

v - speed that the object is moving

c - speed of light (3×10^8 m/s)

m_0 - mass of moving object

m - mass appeared to a stationary observer

Example 16.1

The Enterprise has a length of 650 meter, it can travel at the speed of $0.8V_c$ under the condition that its maximum mass should not exceed 30,000 kg.

a) calculate the observed length at its maximum speed

b) calculate the observed mass at its maximum speed

Solution: a) $\quad L = L_0 \times \sqrt{1-(v/c)^2}$

$\quad\quad\quad\quad\quad L = 650 \times \sqrt{1-(0.8c/c)^2}$

$\quad\quad\quad\quad\quad L = 390$ meter

The length observed by the observer would be 390 meter

b) $\quad m = m_0 \div \sqrt{1-(v/c)^2}$

$\quad\quad\quad m = 30000 \div \sqrt{1-(0.8/c)^2}$

$\quad\quad\quad m = 50{,}000$ kg

The mass measured by the observer would be 50,000 kg

According to Einstein we can conclude that gravity has no effect on mass, time or length if we are moving at the speed relative to speed of light. Hence, we cannot travel at the speed of light as long as we have a mass. The only two factors that matter most in space are time and space (dimension). Muon are particles that support the postulation of Einstein.

Relativistic energy

Total Energy = Rest Energy + Kinetic Energy

$$E = mc^2 + (\Upsilon - 1)mc^2$$

$$E = \Upsilon(mc^2)$$

The relativistic energy expression $E = mc^2$ is a statement about the energy an object contains as a result of its mass and is not to be construed as an exception to the principle of conservation of energy. Energy can exist in many forms, and mass energy can be considered to be one of those forms. At the same time mass can be converted into energy and vice versa.

$$\Delta E = \Delta m\, c^2$$

(Joules)   (Kg)   (3×10^8 m/s)

Practice
Chapter 16

1. What is the relative velocity of light shot from a space-craft moving with a velocity of 2×10^8 m/s ?

    a) 2×10^8 m/s

    b) 3×10^8 m/s

    c) 4×10^8 m/s

    d) 5×10^8 m/s

    e) 6×10^8 m/s

2. Which of the following is TRUE about Einstein postulation ?

    a) electrons can travel at the speed of light

    b) protons can travel at the speed of light

    c) neutrons can travel at the speed of light

    d) photons can travel at the speed of light

    e) all of them are true

3. What is the rest energy of an electron ?

    a) 1.6×10^{-19} J

    b) 1.6×10^{-16} J

    c) 2.6×10^{-15} J

    d) 7.3×10^{-15} J

    e) 8.2×10^{-14} J

4. If a Jedi travels with his battle-cruiser at for 3 years in space at the speed of 0.4c, how long would it be on Earth ?

   a) 3.27 years

   b) 6.87 years

   c) 42.7 years

   d) 70.7 years

   e) 213 years

Answer: Chapter 16

Question	Answer	Explanation
1.	B	According to Einstein light will not travel faster than 3×10^8 m/s So our $V_{relative}$ is still 3×10^8 m/s (speed of light in vacuum)
2.	D	Light is a vibration of photons so they must travel at the speed of light. Photons are nearly massless (10^{-50} kg)
3.	E	$E = m_{electron}\ c^2$ $E = (9.11 \times 10^{-31})(3 \times 10^8)^2$ $E = 8.2 \times 10^{-14}$ J
4.	A	$T = t_0 \cdot \Upsilon = \dfrac{t_0}{\sqrt{1-(v/c)^2}}$ $T = \dfrac{3}{\sqrt{1-(0.4c/c)^2}}$ $T = 3.27$ years

Practice - Test

Question 1 -5 refer to the following choices:

(a) Frequency

(b) Pressure

(c) Energy

(d) Magnetic flux

(e) Power

1. N·m
2. N m^{-2}
3. V·s
4. s^{-1}
5. J s^{-1}

Question 6 - 10 refer to the following law:

(a) Law of conservation of energy

(b) Law of conservation of momentum

(c) Law of conservation of torque

(d) Ohm's Law

(e) Ideal gas law

6. Used to calculate volume of a heated balloon

7. Used to calculate velocity of a baseball after hit by a bat

8. To find height achieved by a ball thrown up vertically

9. Can be used to find the weight required to balance the see-saw

10. Is used to calculate the p.d. across a bulb

Question 11 - 15 use the following graph

a)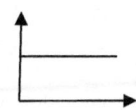

b)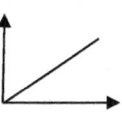

c)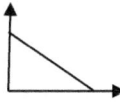

d)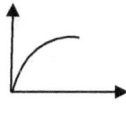

e)

11. The graph of displacement versus time of a train moving at a constant velocity

12. The graph of speed versus time of a ball falling down to the ground

13. The graph of Boyle's law

14. The graph of horizontal velocity vs time of a projectile

15. The graph of p.d. vs current that of a filament bulb

16. How long approximately will it take an object of 5 kilogram to fall vertically down from a 50 meter tall building ?

a)　　1.2 s

b)　　1.6 s

c)　　2.5 s

d)　　3.2 s

e)　　5.1 s

Question 17 - 19 use the information below:

A projectile is launched horizontally with a velocity of 30 m/s at the angle of 45°. Given that there is no air resistance.

17. How high will the object rise vertically ?

a)   10.6 m

b)   22.5 m

c)   30.7 m

d)   43.2 m

e)   59.1 m

18. How long does the journey take ?

a)   1.06 s

b)   2.12 s

c)   3.18 s

d)   4.24 s

e)   4.92 s

19. How far does the object travel horizontally?

a)   30 m

b)   45 m

c)   50 m

d)   75 m

e)   90 m

Question 20 - 21 use the information below:

London-eye has a radius of 60 m took about 20 second to complete one revolution.

20. What is its angular velocity ?

a)   0.144 rad/s

b)   0.223 rad/s

c)   0.314 rad/s

d)   0.413 rad/s

e)   0.523 rad/s

21. What is the centripetal acceleration experienced by the passenger ?

a)   1.09 ms^{-2}

b)   2.51 ms^{-2}

c)   3.04 ms^{-2}

d)   4.32 ms^{-2}

e)   5.91 ms^{-2}

22. A system undergoes isothermal expansion, 150 J of heat is supplied to the system. How much work is done on the system that expand from 30 cm³ to 45 cm³ ?

a)   0   J
b)   75   J
c)   150 J
d)   300 J
e)   2250 J

Question 23 - 24

A canon-gun toy of mass 300g shoots a canon-ball of mass 20g out with a speed of 5m/s. Assuming there is no heat loss.

23. What is the recoil velocity of the canon ?

a)   0.33   m/s
b)   0.66   m/s
c)   1.33   m/s
d)   3.25   m/s
e)   5   m/s

24. If the gun and the ball were in contact for 0.02 second, then what is the force of impact on the ball ?

a)   5   N
b)   50   N
c)   500   N
d)   5000   N
e)   10000 N

25. Temperature of 50K is equivalent to

a)   323 °C
b)   223 °C
c)   50 °C
d)   -50 °C
e)   -223 °C

26. A D.J. at 60 cent club decided to raise the loudness of the sound up, which of the property(s) of sound wave changes?

a)   frequency
b)   amplitude
c)   quality
d)   pitch and quality
e)   amplitude and pitch

27. A uniform rod of mass 4 kilogram is in equilibrium due to the weight hung at point P. What is the value of the weight ?

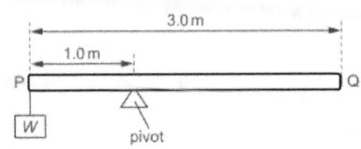

a) 10 N
b) 20 N
c) 30 N
d) 40 N
e) 80 N

28. Magnetic field will not have any effect on which particle ?

a) proton
b) electron
c) photon
d) alpha particles
e) beta particles

29. An object that is white shinny are generally good

a) absorber of heat
b) refractor of heat
c) reflector of heat
d) radiator of heat
e) all of the above

30. An object is placed 5 cm from the mirror, which of the following is the correct properties of the image produced ?

a) real and inverted
b) virtual and inverted
c) real and 5 cm away from the object
d) virtual and 5 cm away from the object
e) virtual and 10 cm away from the object

31. Which of the following wave would travel slowest in air ?

a) sound wave
b) infra-red
c) radio wave
d) beta ray
e) gamma ray

Question 32 - 34 use the information below

A block of mass 0.5 kg is compressed against the spring with a spring constant of 3 N/m.

32. What is the frequency of the oscillation ?

a)   0.25   Hz

b)   0.39   Hz

c)   0.86   Hz

d)   1.79   Hz

e)   2.56   Hz

33. If the block is released from the spring after a compression of 10 cm at what velocity will it move ?

a)   9.2   cm/s

b)   10.0   cm/s

c)   12.5   cm/s

d)   21.4   cm/s

e)   24.5   cm/s

34. The block that was released collide with the black ball of mass 0.2kg, the speed of the block after collision is 8 cm/s. What is the velocity of the black ball ?

a)   21.4   cm/s

b)   24.5   cm/s

c)   32.3   cm/s

d)   37.1   cm/s

e)   41.3   cm/s

35. A bus starting from rest with a uniform acceleration of 8 m/s^2 for 5 seconds in a straight road would cover a distance of

a)   20   m

b)   40   m

c)   80   m

d)   100   m

e)   200   m

36. An object moving at constant speed will

a)   have no acceleration

b)   have a positive resultant force

c)   have a positive acceleration

d)   have a negative resultant force

e)   have a negative acceleration

Question 37 - 40 use below information

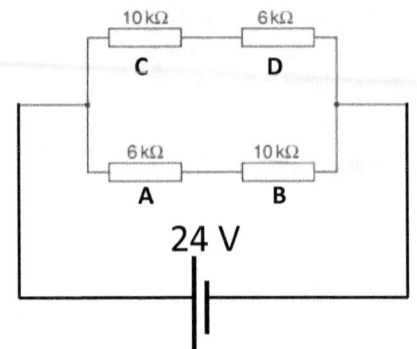

37. What is the total resistance of the circuit ?

a) 4 kΩ
b) 8 kΩ
c) 16 kΩ
d) 32 kΩ
e) 48 kΩ

38. What is the potential difference at resistor A ?

a) 9 V
b) 12 V
c) 15 V
d) 20 V
e) 24 V

39. What is the power supply by the battery ?

a) 0.003 W
b) 0.012 W
c) 0.036 W
d) 0.048 W
e) 0.072 W

40. If we remove resistor A and C from the circuit what would the total resistance of the circuit be ?

a) 2.5 kΩ
b) 3.75 kΩ
c) 5.5 kΩ
d) 8.0 kΩ
e) 16.0 kΩ

41. In what way does isotopes of an atom differs from one another ?

a) different charges
b) different number of electrons
c) different number of neutrons
d) different number of protons
e) different number of photons

SAT Physics (Physics made simple)

42. Which quantity would result from a calculation in which a potential difference is multiplied by an electric charge?

a) electric current

b) electric energy

c) electrical resistance

d) electric field strength

e) electric power

Question 43 - 45 use the information below

Visible light enters the glass that has a refractive index of 1.52

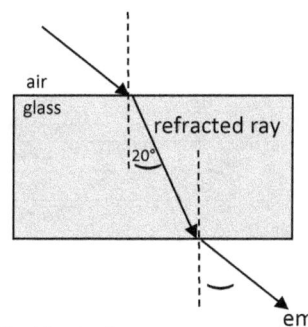

43. What is the angle of incidence ?

    a) 13.0°

    b) 24.3°

    c) 25.6°

    d) 31.3°

    e) 42.5°

44. What is the speed of light in this glass ?

    a) 1.97×10^8 m/s

    b) 2.32×10^8 m/s

    c) 3×10^8 m/s

    d) 4.56×10^8 m/s

    e) 5.06×10^8 m/s

45. What is the critical angle of glass block ?

    a) 23.0°

    b) 34.3°

    c) 35.6°

    d) 38.3°

    e) 41.1°

Question 46 - 48 use the graph below

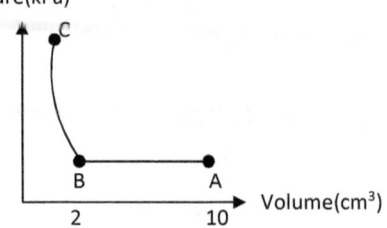

46. A box contain air of volume 10 cm³ at 1 atm is compressed to 2 cm³. How much work is done in compressing the air ?

a) 0.65 J
b) 0.81 J
c) 3.50 J
d) 6.28 J
e) 8.00 J

47. What process is taking place from B to C ?

a) Isothermal
b) Isobaric
c) Isochoric
d) Adiabatic
e) Radioactive

48. Given that from B to C work done in compressing the gas further is 0.35 J. What is the change in internal energy ?

a) - 0.35 J
b) - 1.00 J
c) - 1.116 J
d) 0.85 J
e) 0.35 J

49. What does 'p' and 'q' measures ?

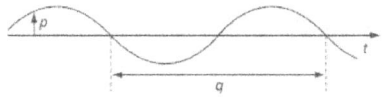

	p	q
a)	Amplitude	Period
b)	Amplitude	Wavelength
c)	Displacement	Period
d)	Displacement	Wavelength
e)	Amplitude	Frequency

50. How much energy is need to melt 500 grams of ice ?

a) 167 kJ
b) 334 kJ
c) 668 kJ
d) 986 kJ
e) 1670 kJ

51. A 20 W light bulb has a charge of 200C flows through it in 50 seconds time. What is the potential difference across the bulb ?

a) 0.4 V

b) 0.5 V

c) 4 V

d) 5 V

e) 10 V

52. Two silver wires A and B have the same area. Wire A is four times longer than wire B. What is the ratio of the resistance of wire B to wire A ?

a)   1/2

b)   1/4

c)   1/8

d)   2

e)   4

53. After an element A undergoes a beta decay, which of the following is true about its daughter nucleus ?

a) it loses one amu

b) it remains the same

c) it gains one amu

d) it loses two amu

e) it gains two amu

54. Which of the following is/are true about inelastic collision ?

(I)   momentum is conserved

(II)   kinetic energy is conserved

(III)   total energy is conserved

a)   (I) only

b)   (II) only

c)   (I) and (II)

d)   (I) and (III)

e)   (I) , (II) and (III)

55. Which word best describe H-2 and H-3 ?

a)   molecules

b)   cathode

c)   anode

d)   ions

e)   isotopes

56. Which of the following is used in time-delay circuit ?

a)   Diode

b)   Thermistor

c)   LDR

d)   LED

e)   Capacitor

57. Which of the following is not part of electromagnetic wave ?

a) Infra-red rays

b) Beta rays

c) Gamma rays

d) Radio wave

e) X-rays

Question 58 - 60 use information below:

A wave moves thru double slit, given that distance between the slits are 5 cm and the distance from slit to the screen is 20 cm, while distance between the fringes are 2 cm.

58. What is the wavelength of this wave ?

a) 0.5 mm

b) 1.0 mm

c) 2.5 mm

d) 5.0 mm

e) 10 mm

59. What is the distance from central maxima to second brightest fringe ?

a) 2 cm

b) 4 cm

c) 6 cm

d) 8 cm

e) 10 cm

60. If distance between the screen and the slit is double what is the distance between the fringes ?

a) 1 cm

b) 2 cm

c) 3 cm

d) 4 cm

e) 5 cm

61. Greatest constructive interference occurs when two waves are having a phase difference of

a) 30°

b) 90°

c) 180°

d) 270°

e) 360°

SAT Physics (Physics made simple)

Question 62 - 64 use the following information:

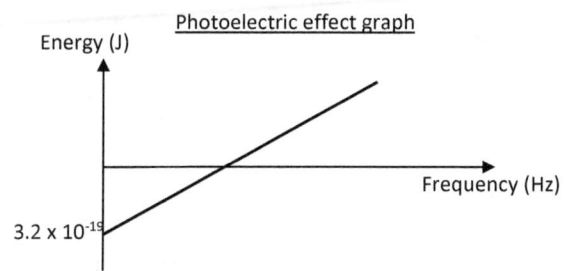
Photoelectric effect graph

62. What is the threshold frequency?

a) 4.83 x 10^{14} Hz
b) 3.24 x 10^{14} Hz
c) 3.05 x 10^{14} Hz
d) 2.07 x 10^{14} Hz
e) 1.53 x 10^{14} Hz

63. What is the maximum kinetic energy of photoelectron if the electromagnetic wave incidence has a wavelength of 420 nm?

a) 3.24 x 10^{-19} J
b) 3.05 x 10^{-19} J
c) 2.07 x 10^{-19} J
d) 1.53 x 10^{-19} J
e) 1.07 x 10^{-19} J

64. According to question 63, what is the speed of emitted photoelectron?

a) 5.81 x 10^{5} m/s
b) 6.12 x 10^{5} m/s
c) 8.91 x 10^{5} m/s
d) 2.31 x 10^{6} m/s
e) 3.00 x 10^{8} m/s

65. What is the period of a wave with a frequency of 5 GHz?

a) 2 ns
b) 20 ns
c) 200 ps
d) 2000 µs
e) 20000 µs

66. One electron-Volt is equivalent to

a) 1.6 x 10^{19} J
b) 1.6 x 10^{-19} J
c) 9.11 x 10^{31} J
d) 9.11 x 10^{-31} J
e) 6.63 x 10^{-34} J

SAT Physics (Physics made simple)

Question 67 - 69 use the information below:

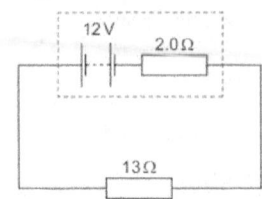

A cell has an e.m.f. of 12 volt is connected to a 13 ohms resistor. The cell has an internal resistance of 2 ohms.

67. What is the potential difference through 13 Ω resistor?

a) 1.6   V
b) 2.4   V
c) 6.0   V
d) 10.4   V
e) 12.0   V

68. What is the rate of energy used by the 13Ω resistor ?

a) 12   W
b) 9.60   W
c) 8.32   W
d) 5.40   W
e) 1.28   W

69. How much power is loss ?

a) 12   W
b) 9.60   W
c) 8.32   W
d) 5.40   W
e) 1.28   W

70. If a metal ring is heated then which of the following quantity would decreases ?

a) density
b) diameter
c) thickness
d) brightness
e) volume

71. How many electrons orbit the hydrogen isotopes H-3 ?

a) 0
b) 1
c) 2
d) 3
e) 4

Question 72 - 75 refer to the diagram below:

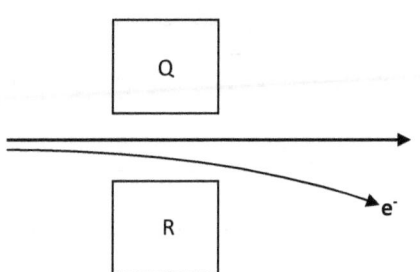

Two rays pass through a region of electric field Q-R. One is unknown and another is a ray composed of a high speed electron.

72. Which kind of radiation is ray that deflect towards the pole R ?

a) alpha
b) beta
c) x-rays
d) gamma
e) electromagnetic ray

73. Which kind of radiation is the ray that moves thru without any deflection called ?

a) alpha
b) beta
c) proton beam
d) gamma
e) cathode ray

74. Which of the following statement is true ?

a) Pole 'R' is positively charged
b) Pole 'R' is negatively charged
c) Pole 'Q' is negatively charged
d) Pole 'Q' has no charge
e) Pole ' Q' and 'R' have the same charge

75. What change would take place(if any) if we shoot a positron through the region instead of an electron ?

a) no change
b) more deflection in same direction
c) no deflection occur
d) same deflection in opposite direction
e) more deflection in opposite direction

Answer Key

Question	Answer	Explanation
1.	C	Nm → force x distance = Work or Energy
2.	B	Nm^{-2} → force ÷ area = Pressure
3.	D	Vs → p.d. x time = Magnetic flux
4.	A	s^{-1} → 1/Time = Frequency
5.	E	Js^{-1} → energy ÷ time = Power
6.	E	We can apply the concept of ideal gas here $V_1 : T_1 = V_2 : T_2$
7.	B	Using law of conservation of momentum $m_1u_1 + m_2u_2 = m_1v_1 + m_2v_2$
8.	A	We can use law of conservation of energy here KE = PE
9.	C	Turning effect of force in equilibrium can be view as Total clockwise torque = Total anti-clockwise torque
10.	D	Ohm's Law can be used to help calculate the p.d. V = I R
11.	B	The graph of displacement vs time of an object moving at a constant velocity means that we can view this as a directly proportional graph As time increases the displacement also increases
12.	B	As the ball fall its velocity is zero (starts from rest) then velocity is incresing due to the effect of gravity. We can say graph of speed vs time is directly proportional
13.	E	Boyle's law say Presurre is inversly proportional to volume
14.	A	Horizontal velocity of a projectile always remain constant
15.	D	The graph is almost directly proportional, except that filament bulb produces heat so its going to curve due to heat loss

Question	Answer	Explanation
16.	D	$Y = ut + 0.5 at^2$ $50 = 0(t) + 0.5(10)t^2$ $T = 3.16 \approx 3.2$ sec
17.	B	$V^2 = U^2 + 2as$ $0^2 = (30\sin 45)^2 + 2(10)s$ $s = 22.5$ m
18.	D	$T_{up} = V_i \sin\theta \div g$ $T_{up} = 30 \sin 45 \div g$ $T_{up} = 2.12$ s $T_{down} = T_{up}$ Total time $= T_{up} + T_{down}$ Total time $= 2.12 + 2.12 = 4.24$ second
19.	E	Range $= V_i \cos\theta \times 2 t_{up}$ Range $= 30 \cos 45 \times 2(2.12)$ Range $= 90$ meter
20.	C	$\omega = \Delta\theta \div \Delta t$ $\omega = 2\pi \div 20$ $\omega = 0.314$ rad/s
21.	E	$a_c = r \times \omega^2 = 60 \times (0.314)^2 = 5.91$ ms^{-2}
22.	C	During isothermal expansion $\Delta U \rightarrow Q = W$ $Q = 150$ J $W = 150$ Joules
23.	A	total momentum before $=$ total momentum after $0 = 300(v) + 20(5)$ $v = 0.333$ m/s
24.	A	$\Delta P = m \Delta v$ $F\Delta t = 0.02(5-0)$ $F \times 0.02 = 0.1$ $F = 5$ N
25.	E	$T_c = T_k - 273$ $T_c = 50 - 273 = -223$ °C
26.	B	Loudness of sound depend mainly on the amplitude

Question	Answer	Explanation
27.	B	In equilibrium Total clockwise torque = Total anti-clockwise torque Weight of rod x D_1 = W x D_2 40 x 0.5 = W x 1 W = 20 N
28.	C	A photon has no mass and no charge, therefore it will not experience any field effect.
29.	C	White shinny object are a good reflector of heat radiation, since white is the color that absorb very little heat.
30.	E	An object 5 cm from the mirror will have a virtual image of 5 cm inside the mirror. So we can say that the image is virtual and 10cm away from the object.
31.	A	Choice b,c and e are electromagnetic wave which travels at the speed of light Choice d travels nearly equal to speed of light Sound wave would travel slowest (330 m/s)
32.	B	We can find period first by using $$T = 2\pi\sqrt{\frac{m}{k}} = 2\pi\sqrt{\frac{0.5}{3}} = 2.56 \text{ sec}$$ We want the frequency so $$F = \frac{1}{T} = \frac{1}{2.56} = 0.39 \text{ Hz}$$
33.	E	Elastic PE = Kinetic enegy 0.5 k x^2 = 0.5 m v^2 (3) $(10/100)^2$ = (0.5) v^2 v = 24.5 cm/s
34.	E	Total momentum before = Total momentum after (0.5) (24.5) = (0.5) (8) + 0.2 (v) v = 41.3 m/s
35.	D	We can use equation of motion here s = u t + 0.5 a t^2 s = (0)(5) + 0.5(8)$(5)^2$ s = 100 m
36.	A	Constant speed means that the acceleration is equal to zero

Question	Answer	Explanation
37.	B	$R_{AB} = R_A + R_B = 6k + 10k = 16\ k\Omega$ $R_{CD} = R_C + R_D = 10k + 6k = 16\ k\Omega$ $1/R_{total} = 1/R_{AB} + 1/R_{CD}$ $1/R_{total} = 1/16 + 1/16$ $R_{total} = 8\ k\Omega$
38.	B	$V_A = R_A \times V_{total} / R_{AB}$ $V_A = 6 \times 24 / 16$ $V_A = 9\ V$
39.	E	$I_{total} = V_{total} / R_{total}$ $I_{total} = 24 / (8 \times 10^3)$ $I_{total} = 0.003\ A$ $P_{supply} = I_{total} \times V_{total}$ $P_{supply} = 0.003 \times 24$ $P_{supply} = 0.072\ W$
40.	B	So only D and B is left $\quad\quad\quad$ 6 kΩ $\quad\quad\quad$ 10 kΩ $1/R_{total} = 1/6 + 1/10$ $R_{total} = 3.75\ k\Omega$
41.	C	Isotopes of an atom would have the same number of protons but different number of neutrons
42.	B	If we multiply $V \times Q \rightarrow (I\ R) \times (I\ T)$ $\quad\quad\quad I^2\ R \times T =$ Power x Time = Energy
43.	D	$n_{air}\ \sin\theta_i = n_{glass}\ \sin\theta_r$ $1 \times \sin\theta_i = 1.52\ \sin 20$ $\theta_i = 31.3°$
44.	A	$\quad\quad$ Air $\quad\quad\quad$ Glass $\quad\quad n_1 v_1 = n_2 v_2$ $1 \times (3.10^8) = 1.52 \times V$ $\quad\quad V = 1.97 \times 10^8\ m/s$
45.	E	$n = 1/\sin \theta_c$ $\theta_c = \sin^{-1}(1/n)$ $\theta_c = 41.1°$

Question	Answer	Explanation
46.	B	Work = P × ΔV Work = $(101 \times 10^3) \times [(10-2) \times 10^{-6}]$ Work = 0.81 J
47.	D	The process from B to C is known as adiabatic compression Because no heat was added to the system
48.	A	The process from B to C is adiabatic which means Q = 0 So we apply law of thermodynamic $\quad$ Q $\;= \Delta U + W$ $\quad$ 0 $\;= \Delta U + 0.35$ $\quad \Delta U = -0.35$ J
49.	C	'p' is not the amplitude it is the displacement of the wave 'q' is the time period of wave
50.	A	We use Q = m H_v $\quad$ = 0.5 × 334 kJ/kg = 167 kJ
51.	D	We want to find p.d. V We first find I by using I = ΔQ ÷ ΔT $\quad$ we get I = 200 ÷ 50 = 4 A Then we use $\quad$ P = I V $\quad\quad$ 20 = 4 (V) $\quad\quad$ V = 5 volt
52.	B	To find $R_B : R_A \rightarrow$ We refer to R is directly proportional to L $\quad R_A / L_A = R_B / L_B$ $\quad R_A / 4L_B = R_B / L_B$ $\quad L_B / 4L_B = R_B / R_A$ $\quad R_B : R_A = 1:4$ $\quad R_B : R_A = 1/4$
53.	C	After lossing an electron element A will have +1 to its proton number This increases the amu by 1
54.	D	During an inelastic collision a momentum is conserved but kinetic energy is not Keep in mind that total energy is always conserved, some energy during the collision turned into heat and sound during the impact.
55.	E	Isotopes are shown here since H-2 and H-3 has same number of proton but different number of neutrons

Question	Answer	Explanation
56.	E	In a time-delay circuit we have to use capacitor because we want it to be fully charged first then it wil give discharge , allowing current to flow in the circuit.
57.	B	All of the following are part of EM except Beta rays Beta is made of a high speed electron
58.	D	$\frac{\lambda}{d} = \frac{x}{L} \rightarrow \frac{\lambda}{5} = \frac{2}{20}$ λ = 0.5 cm or 5 mm
59.	B	From n = 0 to n = 2 $\rightarrow$ N = 2 We can use N·x = 2 · 2 = 4 cm
60.	D	Distance L is directly proportional to fringe x We can say $L_1 / x_1 = L_2 / x_2 \rightarrow$ 20 /2 = 40 / x_2 x_2 = 4 cm
61.	E	Greatest or maximum constructive interference occurs when two waves are inphase or phase difference is 0 or 360°
62.	A	To find threshold frequency we set E_k = 0 E_k = hf - Φ 0 = (6.63 x 10^{-34}) f - 3.2 x 10^{-19} f = 4.83 x 10^{14} Hz
63.	D	E_k = hf - Φ E_k = (6.63 x 10^{-34})(3x 10^8)/(420 x 10^{-9}) - 3.2 x 10^{-19} E_k = 1.53 x 10^{-19} J
64.	A	E_k = 1/2 m v^2 $V_{electron}$ = (2 x E_k / $m_{electron}$)$^{1/2}$ $V_{electron}$ = 580645 = 5.81 x 10^5 m/s
65.	C	Frequency = 1/ period Period = 1/ frequency Period = 1/ (5 x 10^9) = 2 x 10^{-10} s or 200 ps
66.	B	1 eV = 1.6 x 10^{-19} J

Question	Answer	Explanation
67.	D	We apply Ohm's Law by adding the two resistor in series → $R_T = 2 + 13 = 15\ \Omega$ $V_T = I_T R_T$ $12 = I_T \times 15$ $I_T = 0.8\ A$ $V_R = I_T \times R = 0.8 \times 13 = 10.4\ V$
68.	C	To find rate of energy or power used by the resistor we use $P = I^2 \times R$ $P = 0.8^2 \times 13$ $P = 8.32\ w$
69.	E	To find rate of energy loss we calculate power loss in internal resistor $P = I^2 \times r$ $P = 0.8^2 \times 2$ $P = 1.28\ w$
70.	A	When we heat an object it would expand, which means more volume Volume is inversely proportional to density → $D = \dfrac{m}{V}$ → more 'v' means less 'D' So density decreases.
71.	B	Isotopes have same number of proton(s) Hydrogen have only 1 proton which means there exist only 1 electron in the orbital shell
72.	B	Beta particles are made of electrons that are moving in high speed
73.	D	Gamma rays are electromagnetic waves that experienced no effect on electric field
74.	A	By looking at the path that electron beam moves we can conclude that 'R' must be postively charged
75.	D	Positrons are postively charged particles which have similar mass to electrons, therefore it would deflect in the same manner but in the opposite direction

SAT Physics (Physics made simple)

Key words

A

Absorber - A substance that takes in radiation
Alpha Radiation - Alpha particles, Each consists of 2 Protons and 2 Neutrons (A Helium Nucleus), emitted by unstable nuclei
Alternating Current - Electric current in a circuit that repeatedly changes direction
Amplitude - The height of a wave crest from trough to the rest position, maximum Distance moved by an object from its equilibrium position
Angle of Incidence - Angle between the incident ray and the normal
Angle of Reflection - Angle between the reflected ray and the normal
Atomic Number - The number of protons in an atom

B

Beta Radiation - High energy electrons that are created in and emitted from unstable nuclei
Biofuel - Fuel made from animal and plant products
Black Hole - An object in space that has so much mass, that nothing, not even light, can escape from its gravitational field

C

Centre of Mass - The point where an object's mass may be thought to be concentrated
Centripetal Acceleration - The acceleration of an object moving in a circle at a constant speed. Centripetal acceleration always acts towards the centre of the circle
Centripetal Force - The resultant force towards the centre of a circle acting on an object moving in a circular path
Chain Reaction - Reactions in which one reaction causes further reactions, which in turn cause further reactions
Chain Reaction of Nuclear Fission - When fission neutrons cause further fission, so more fission neutrons are released. This goes on to produce further fission
Chemical Potential Energy - Energy stored in form of chemical
Circuit Breaker - An electromagnetic switch that opens and cuts the current off if too much current passes through it
Compression - high pressure of particles
Condensation - Turning from a vapour into a liquid
Conduction - Transfer of energy from particle to particle in matter via free electrons
Conductor - Material or Object that conducts heat or electricity

Conservation Of Energy - Energy cannot be created or destroyed (It can only be transferred)
Conservation Of Momentum - In a closed system, the momentum before an event is equal to the total momentum after the event. Momentum is conserved in any collision or explosion provided that no external forces act on the objects that collide or explode.
Convection - Transfer of energy in fluid
Convection Current - The circular motion of matter caused by the heating in fluids
Converging Lens - A lens that makes light rays parallel to the principal axis converge to a point
Convex Lens - Converging Lens
Concave Lens - Diverging Lens
Critical Angle -The angle of incidence of a light ray in a transparent substance which produces refraction
Crumple Zone - Region of a vehicle designed to crumple in a collision to reduce the force on the occupants

D

Diffraction - The spreading of waves when they pass through a gap or around the edges of an obstacle that has a similar size as the wavelength of the waves
Diode - Electrical device that allows current flow in one direction only
Doppler Effect - The change of wavelength (and frequency) of the waves from a moving source due to the motion of the source towards or away from the observer
Drag Force - A force opposing the motion of an object due to fluid (e.g air) flowing past the object as it moves

E

Echo - Reflection of sound that can be heard
Efficiency - Useful Energy transferred by a device ÷ total energy supplied to the device
Effort - The force applied to a device used to raise a weight or shift an object
Elastic - A material is elastic if it is able to regain its shape after it has be squashed or stretched
Elastic Potential Energy - Energy stored in an elastic object when work is done to change its shape
Electrical Energy - Energy transferred by the movement of electrical charge
Electromagnetic Induction - The process of introducing a potential difference in a wire by moving the wire so it cuts along the lines of force of a magnetic field
Electromagnetic Wave - Electric and magnetic disturbance that transfers energy from one place to another
Electron - A tiny particle with a negative charge Electrons orbit the nucleus in atoms or ions.
Emitter - A substance that gives out radiation
Equilibrium - The state of an object when it is at rest
Evaporation - Turning from a liquid to vapour

F

Fluid - A liquid or a gas
Focal Length - The distance from the centre of a lens to the point where light rays parallel to the principal axis are focused
Fossil Fuel - Fuel obtained from long dead biological material
Frequency - The number of wave crests passing a fixed point every second (measures in Hertz)
Friction - Forces opposing the movement of one surface over another
Fuse - Contains a thin wire that melts and cuts the current if too much current is passing through it

G

Gamma Radiation - Electromagnetic Radiation emitted from unstable nuclei in radioactive substances
Gamma Ray - The highest energy eaves in the electromagnetic spectrum
Generator - A machine generates electricity by conversion of energy
Geothermal Energy - Energy from hot underground rocks
Gravitational Attraction - Force that pulls two masses together

H

Half-Life - Average time taken for the number of nuclei of the isotope (or mass of the isotope) in a sample to halve
Hooke's Law - The extension of a spring is directly proportional to the force applied provided the limit is not exceeded
Hydraulic Pressure - The pressure in a liquid in a hydraulic arm

I

Infrared Radiation - Electromagnetic waves between visible light and microwaves in the electromagnetic spectrum
Input Energy - Energy supplied to a machine
Insulating - Reducing energy transfer by heating
Insulator - Material or object that is a poor conductor of heat and electrons
Ion - A charged particle that is produced by the loss or gain of electrons
Ionisation - Any process in which atoms become charged
Isotope - Atom that has the same number of protons but a different number of neutrons. It has the same atomic number but a different mass number

L

LDR (Light Dependent Resistor) - Device with a resistance that varies with the amount of light falling on it
Limit of Proportionality - The limit for Hooke's Law applied to the extension of a stretched spring.
Load - The weight of an object raised by the a device used to lift the object or the force applied by a device to shift an object
Longitudinal Wave - Wave in which the vibrations are parallel to the direction of energy transfer.

M

Machine - A device in which a force applied at a point produces another force at another point
Magnifying Glass - A converging lens used to magnify a small object that must be placed between the lens and the focal point
Mass Number - The total number of protons and neutrons in an atom
Mechanical Wave - Wave that travels through a substance
Microwave - Part of the electromagnetic spectrum

N

National Grid - The network of cables and transformers used to transfer electricity from power stations to consumers
Neutral Wire - The wire of a main circuit that is earthed at the local substation so its potential is close to zero
Neutron - A no charge particle found in the nucleus
Non-renewable - Something that cannot be replaced once it is used up
Normal - Straight line through a surface or boundary perpendicular to it
Nucleus - The very small and dense central part of an atom composed of proton and neutron

O

Ohm's Law - The current through a resistor is directly proportional to the potential difference across a resistor
Ohmic Conductor - A conductor that has a constant resistance and therefore obeys Ohm's law
Optical Fiber - Thin glass fiber used to send light signals along
Oscillation - movement of wave in particular direction

P

Parallel - Components connected in a circuit so that the potential difference is the same across each one
Pitch - The frequency of a sound wave
Potential Difference - A measure of the work done or energy transferred per coulomb of charge.
Power - The amount energy transferred per second
Pressure - Force per unit area for a force acting on a surface at right angles to the surface
Principal Axis - A straight line that passes along the normal at the centre of each lens surface
Principal Focus - The point where light rays parallel to the principal axis are focused
Proton - A positively charged particle found inside the nucleus of an atom

R

Radio wave - Longest wavelength of the electromagnetic spectrum
Real Image - An image that forms where light rays meet and can be projected on the screen
Reflector - A surface that reflects radiation
Refraction - The change of direction of a light ray when it passes across a boundary between two transparent substances
Renewable Energy - Energy from sources that never run out
Resistance - Anything that slows the flow of current in a circuit, measures in Ohms
Resultant Force - The combined effect of the forces acting on an object

S

Series - Components connected in a circuit so that the same current that passes through them are in series with each other
Solar Cell - Electrical Cell that produces a voltage when in sunlight. Usually connected together in solar cell panels
Solar Energy - Energy from the sun
Sound - A form of mechanical energy
Specific Heat Capacity - Energy needed by 1kg to raise its temperature by 1°c
Step-Down Transformer - Electrical Device that is used to step down the size of an alternating voltage
Step-Up Transformer - Electrical Device that is used to step up the size of an alternating voltage

T

Temperature - The degree of hotness or coldness of a substance
Terminal Velocity - The velocity reached by an object when the drag force on it is equal and opposite to the force making it move
Thermistor - Device with a resistance that varies with temperature
Total Internal Reflection - When the angle of incidence of a light ray in a transparent substance is greater than the critical angle. This means that the angle of reflection is equal to the angle of incidence
Transformer - Electrical device used to change a voltage
Transverse Waves - Waves in which the vibrations are perpendicular to the direction of energy transfer

U

Ultrasound Wave - Sound wave at a frequency greater than 20 000 Hz
Ultraviolet Radiation - Electromagnetic radiation just beyond the blue end of the visible spectrum

V

Virtual Image - An image seen in a lens or a mirror from which light rays appear to come after being refracted by the lens or reflected on a mirror. It cannot be projected

W

Wasted Energy - Energy that is not usefully transferred
Wave - a method of transferring energy
Wavelength - The distance from one wave crest to the next wave crest (along the waves)
Work - amount of energy transferred or changed

X

X-Ray - High energy wave from the part of the electromagnetic spectrum between gamma and ultraviolet waves

Appendix I

Units Conversion

Prefixes	Value	Standard form	Symbol
Tera	1,000,000,000,000	10^{12}	T
Giga	1,000,000,000	10^{9}	G
Mega	1,000,000	10^{6}	M
Kilo	1,000	10^{3}	k
deci	0.1	10^{-1}	d
centi	0.01	10^{-2}	c
milli	0.001	10^{-3}	m
micro	0.000001	10^{-6}	µ
nano	0.000000001	10^{-9}	n
pico	0.000000000001	10^{-12}	p

Rules for trigonometry

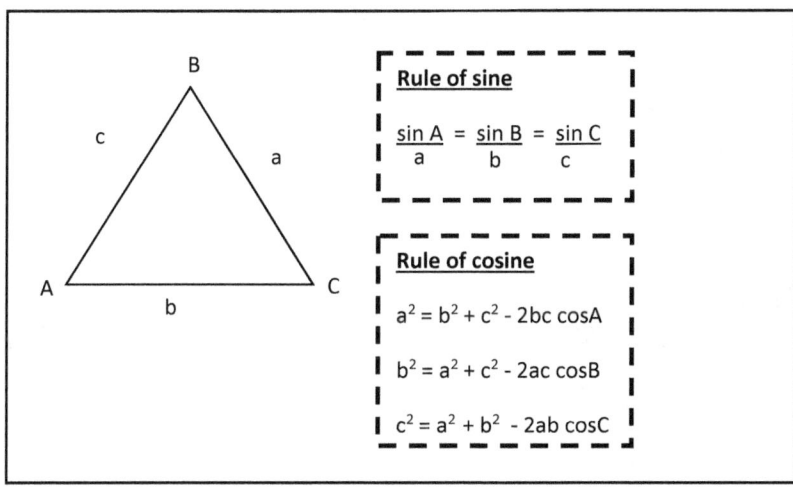

Rule of sine

$$\frac{\sin A}{a} = \frac{\sin B}{b} = \frac{\sin C}{c}$$

Rule of cosine

$$a^2 = b^2 + c^2 - 2bc \cos A$$

$$b^2 = a^2 + c^2 - 2ac \cos B$$

$$c^2 = a^2 + b^2 - 2ab \cos C$$

Appendix II

Mechanics

speed $= \dfrac{\text{change in distance}}{\text{change in time}}$

velocity $= \dfrac{\text{change in displacement}}{\text{change in time}}$

Average speed $= \dfrac{\text{total distance}}{\text{total time}}$

Average velocity $= \dfrac{\text{total displacement}}{\text{total time}}$

Acceleration $= \dfrac{\text{change in velocity}}{\text{change in time}}$

Equation of motion

	Missing term
$v = u + at$	s
$s = ut + 1/2\, at^2$	v
$v^2 = u^2 + 2as$	t
$s = \dfrac{(u+v)\, t}{2}$	a

circular motion

$s = \theta R$

$v = \omega R$

$\omega = 2\pi f$

$a = \alpha R$

Forces

$F = ma$

$f = \mu N$

$W = mg$

$F = kX$

$F_c = m\, a_c$

$a_c = \dfrac{v^2}{R}$

Energy

Work = force x distance

$KE = \dfrac{1}{2} m v^2$

$GPE = mgh$

$EPE = \dfrac{1}{2} k X^2$

Power $= \dfrac{\text{Work}}{\Delta \text{time}} = \dfrac{\Delta \text{energy}}{\Delta \text{time}}$

Momentum

$p = m \cdot v$

$\Delta p = m \cdot \Delta v$

$\Delta p = F \cdot \Delta t$

Torque

$\tau = F \cdot d$

In equilibrium

$\Sigma \tau_{acw} = \Sigma \tau_{cw}$

$\Delta L = m \times \Delta v \times r$

Thermal Properties

Pressure

$P = \dfrac{F}{A}$

$P = h \, \rho \, g$

$PV = nRT$

Thermal expansion

$\Delta L = \alpha \, L_i \, \Delta T$

$\Delta V = \beta \, V_i \, \Delta T$

Heat energy

$Q = m \times c \times \Delta T$

$Q_f = m \times H_f$

$Q_v = m \times H_v$

$Q = \Delta U + W$

Waves

$v = f \cdot \lambda$

$f = \dfrac{1}{T}$

$v = \sqrt{\dfrac{T}{\mu}}$

Beats = |Frequency$_1$ - Frequency$_2$|

$\text{Intensity} = \dfrac{1}{distance^2}$

angle of incidence = angle of reflection

Snell's Law

$n_1 \sin(\theta_1) = n_2 \sin(\theta_2)$

$n = \dfrac{1}{\sin \theta_c}$

Double slit

$\dfrac{\lambda}{D} = \dfrac{X}{L}$

Mirror and Lens

$M = \dfrac{d_i}{d_o} = \dfrac{h_i}{h_o}$

$\dfrac{1}{f} = \dfrac{1}{d_i} + \dfrac{1}{d_o}$

$\dfrac{1}{f} = \dfrac{1}{h_i} + \dfrac{1}{h_o}$

SHM

$T = 2\pi \sqrt{\dfrac{L}{g}}$

$T = 2\pi \sqrt{\dfrac{m}{k}}$

Field and Potentials

Gravitaional field

$K_s = \dfrac{R^3}{T^2}$

$F_g = \dfrac{G\, m_1\, m_2}{R^2}$

$a_g = \dfrac{G\, m}{R^2}$

$\Delta U = -\dfrac{G m_{object} \cdot m_{Earth}}{R}$

$V_{escape} = \sqrt{\dfrac{2GMe}{Re}}$

$V_{orbit} = \sqrt{\dfrac{GMe}{Re}}$

Electric field

$F_e = \dfrac{K\, q_1\, q_2}{R^2}$

$E = \dfrac{kq}{R^2}$

$V = E\, D$

$U = q\, V$

Electricity

$V = I \cdot R$

$P = I \cdot V$

$I = \dfrac{\Delta Q}{\Delta t}$

$\varepsilon = IR + Ir$

$R = \dfrac{\rho L}{A}$

$Q = C\, V$

$U = \dfrac{Q\, V}{2}$

Magnetism

$B = \dfrac{\mu_0\, I}{2\pi R}$

$F = B\, I\, L$

$F = q\, v\, B$

$\Phi = B \cdot A$

Electromagnetic induction

$P_{input} = P_{output}$

$V_{input} \times I_{input} = V_{output} \times I_{output}$

SAT Physics (Physics made simple)

Quantum

$E = hf$

$E_K = hf - \Phi$

$p = \dfrac{h}{\lambda}$

$E = mc^2$

$A_n = \dfrac{A_0}{2^n}$

$M_n = \dfrac{M_0}{2^n}$

$T = t_0 \cdot Y = \dfrac{t_0}{\sqrt{1-(v/c)^2}}$

$L = L_0 \div Y = L_0 \cdot \sqrt{1-(v/c)^2}$

$m = m_0 \cdot Y = \dfrac{m_0}{\sqrt{1-(v/c)^2}}$

Appendix III

Fundamental Constant

Quantity	Symbol	Value
Acceleration due to gravity	g	9.81 m/s² ≈ 10 m/s²
Gravitational constant	G	6.67×10^{-11} Nm² / kg²
Gas constant	R	8.31 J K⁻¹ mol⁻¹
Coulomb constant	k	9×10^9 N m² C⁻²
Permeability of free space	μ_0	$4\pi \times 10^{-7}$ T m A⁻¹
Planck's constant	h	6.63×10^{-34} J s
Speed of light (in Air and Vacuum)	c	3×10^8 m/s
Charge on Proton and electron	e	1.6×10^{-19} C
Mass of electron	m_e	9.11×10^{-31} kg
Mass of proton	m_p	1.673×10^{-27} kg
Mass of neutron	M_n	1.675×10^{-27} kg
Electron-Volt	1eV	1.6×10^{-19} J

www.ingramcontent.com/pod-product-compliance
Lightning Source LLC
Chambersburg PA
CBHW061435180526
45170CB00004B/1421